The U
Rabbit Hole:
Book One

Kelly Chase

Copyright © 2022 Kelly Chase

All rights reserved.

ISBN: 9798354000616

DEDICATION

Written with deep love and reverence for the oneness that connects us all.
May it awaken.

CONTENTS

1	The Pentagon Says That UFOs Are Real.	1
2	Are UFOs Human Technology?	16
3	Are UFOs Alien Technology?	34
4	Are UFOs Humans From The Future?	65
5	The Interdimensional & Ultraterrestrial Hypotheses	77
6	Mr. Delonge Goes To Washington [Part One]	108
7	Mr. Delonge Goes To Washington [Part Two]	128
8	A Rational Approach To Ancient Aliens [Part 1]: Archeology & Epistemology	146
9	A Rational Approach To Ancient Aliens [Part 2]: The Secret History Of Human Civilization	172
10	A Rational Approach To Ancient Aliens [Part 3]: Cataclysms & Non-Human Intervention	200

ACKNOWLEDGMENTS

I would like to acknowledge how deeply strange all of this is.

1 The Pentagon Says The UFOs Are Real.

Have you ever seen a UFO?

I have. Once. It was on summer vacation when I was a teenager. My family was staying in a beach house near Hatteras on North Carolina's Outer Banks. I was sitting out on the deck one night looking out at the vast, black expanse of the ocean and the impossible field of stars above.

Growing up under the dull, orange glow of a suburban night sky, I so rarely got to see stars like that. And so I spent most nights out on the deck in a weathered adirondack chair, with my feet kicked up on the railing, and my head tipped back just marveling at the sheer number of stars. On a clear enough night, you could see that the stars weren't just uniform white pinpricks of light—but rather, they sparkled like gems in every color of the rainbow.

It was on one of those nights that I saw it. At first I thought it was a plane, but the clear, bright light high above me in the sky was moving too fast. My brain reshuffled, trying to make sense of what I was seeing. I knew there was an Air Force base nearby—that made sense. But then the light took a hard, 90-degree left turn without slowing down followed immediately by another equally impossible turn to the right before accelerating off past the horizon, streaking like a comet.

Electrified by what I had seen, I ran inside and told my family, but they laughed at me and then accused me of smoking weed. I didn't talk about it again.

I didn't see anything else on that trip—or ever again in the years since then. I would sometimes think about that clear, bright light with a detached, shapeless curiosity. But what do you do with an experience that you can't attach to anything solid—that falls outside the boundaries of everything you know to be true? I think most people who find themselves in that situation do exactly what I did. I mostly forgot.

While the Kelly of my youth held a worldview expansive enough to permit for a bit of mystery, as I got older something changed. Or rather, I did. A series of life-changing events—including the sudden and devastating loss of my father when I was 20—caused me to forcibly amputate any notion of the spiritual, paranormal, or otherworldly. Life, as I saw it, was a harsh, rigid, unrelenting landscaping painted only in shades of black and white.

Everything could be known—it had to be. Because if it could be known it could be planned for, mitigated, dealt with, or even choked on. But at least I'd see it coming, and I wouldn't make the mistake of hoping it could be different.

There wasn't any room for ambiguity in the paradigm that I adopted. I thought about what I'd seen in the sky that night less and less, and then barely at all until eventually years would go by without that troublesome, incongruent little thought bubbling to the surface

But then in April of last year, something crazy happened. It might even be the craziest thing that's *ever* happened—which is saying a lot considering it coincided with the outbreak of a generational pandemic.

On April 27th, 2020, the Pentagon released a statement to the press confirming that three leaked videos of UFOs that had been included in a 2017 exposé in *The New York Times* weren't just real, but they admitted that they had no idea what these craft were, and had no explanation for their seemingly impossible maneuvering.

We'll get into the whole backstory behind the leaked videos, the *New York Times* article, and this stunning admission from the Pentagon in future chapters. The utterly bizarre and unlikely series of events that ties these things together is at the very heart of this story. But first let's talk about these videos.

Because—*dude.*

So these videos that were leaked in 2017, and then confirmed to be authentic by the Pentagon in 2020, are absolutely mind-blowing—and not because of the quality of the video or what they show. In this respect they are about as grainy and difficult to decipher as any other sketchy UFO video that you can find on YouTube.

What makes these three videos remarkable is that they were filmed by U.S. Navy fighter pilots with the USS Nimitz off the west coast and with the USS Theodore Roosevelt off the east coast. Both of these ships are giant aircraft carriers which are part of huge battle cruiser fleets. This means that there were tons of ships and planes on the scene with the very best instruments available. These videos don't just have multiple eye-witnesses who have been willing to step forward to verify their authenticity, but they have verification of the same object being tracked by multiple sensors both in the air and on the sea. And they each have comprehensive chain-of-custody documentation, which means they can't have been altered.

Basically, you don't have to believe that these videos are real, but if these videos aren't enough to convince you, I doubt anything short of a personal encounter would.

And for the record, if that's how you feel, that's totally valid and I respect it. If anything, when it comes to this particular topic, it is best to proceed cautiously and be skeptical of the evidence. I only bring this up to say that these videos are the closest thing to a smoking gun that we're going to get short of flying saucers landing on the White House lawn. It would be hard to dream up more bulletproof evidence.

So what do these videos show?

The first video is referred to as *Gimbal*. A Gimbal is "a pivoted support that permits rotation of an object about an axis", which makes a lot of sense once you see how this thing moves.

The Gimbal video was captured by U.S. Navy F/A-18 Super Hornet aircrafts utilizing a Raytheon Advanced Targeting Forward-Looking Infrared (ATFLIR) pod. You don't have to know what any of that means, you just need to know that we're talking about some of the most advanced sensors and powerful tracking systems on the market.

In this video, Navy pilots can be heard marveling at a top-shaped object that they first think might be a drone when suddenly it rotates a full 90 degrees while staying in place—something that should be aerodynamically impossible—leaving them baffled as to what they are seeing. Although only one of these objects can be seen in the video, eerily, one of the voices can be heard saying, "There's a whole fleet of them."

The second video, from an incident reported by pilots aboard the USS Nimitz, is called "FLIR" for the advanced ATFLIR system that was used to record it. There is no audio to this video, but in it, you can see a large, smooth, oblong, Tic Tac-shaped object hovering in the sky, unmoving. It just hangs there for a couple of minutes and then suddenly shoots off to the left at a velocity that can only be called *unprecedented*. We don't have anything that could go that fast, and if we did, the laws of physics would mean that anyone onboard would be dead from the g-forces.

The third video is called "Go Fast"—I'll give you three guesses why. This video was filmed by US Navy pilots from the USS Theodore Roosevelt in 2015 and it shows a round or oblong craft flying in the air over the open ocean at an impossible speed. The pilots are laughing and shouting, clearly stunned at what they are seeing. One of the voices you can hear is that of Commander David Fravor who told ABC news in an interview about the event, "After 18 years of flying, I've seen pretty much about everything that I can see in that realm, and this was nothing close."

In the time since, these incidents have been investigated more deeply, with multiple eye-witnesses coming forward to share what they saw, and the details that they have been filling in are even more incredible than what was shown in the videos. For example, during the 2004 Nimitz incident it's reported that sensors aboard the USS Princeton, another aircraft carrier in the area that also encountered the TicTac-shaped craft, captured this craft dropping from 80,000 feet in the air down to the surface of the ocean in 0.78 seconds. That's hundreds of thousands of kilometers per hour. For comparison, our fastest planes can go about Mach 3.2 which is about 4000 kph.

We'll get more into the startling incidents behind these videos as we move forward and explore what these things could possibly be and what it all might mean, but for now what you need to know is simply this:

The Pentagon has confirmed that there are what appear to be intelligently controlled craft that fly in our airspace and interact regularly with some of the deadliest and most advanced arms of our military apparatus with complete impunity. The government says that they don't know what they are. They say that they aren't ours, and that we're pretty sure that they don't belong to any of our adversaries—specifically Russia or China—because the technology being displayed is just simply too advanced for that to really be plausible.

We aren't just talking about a quantum leap forward in technological ability. We're talking about technology that implies an

entirely different paradigm and understanding of the laws of science as we currently know them.

So what exactly do I mean by that? What are we dealing with here?

There's this guy, Luis Elizondo. Lue is the former director of a secretive government agency called the Advanced Aerospace Thread Identification Program (AATIP) that studied UFOs which was first revealed to exist in the 2017 *New York Times* article. Lue is the *former* director of that program because he resigned from his position in 2017 in protest citing "excessive secrecy and internal opposition" in his resignation letter to Defense Secretary James Mattis asking, "Why aren't we spending more time and effort on this issue?"

Yeah, we're going to be talking a lot more about Lue.

Lue Elizondo has spent the last 5 years since his resignation furthering the study of UFOs (or Unidentified Aerial Phenomena as he prefers to call them) and he cites five "observables" of UAP technology:

Observable 1: Anti-Gravity

These craft are somehow able to overcome Earth's gravity with no visible means of propulsion, and no visible flight surfaces such as wings.

Observable 2: Sudden & Instantaneous Acceleration

Like in the first Nimitz video where the TicTac took off to the left at 3600 mph, these craft can go from a standstill to seemingly impossible speeds just like that.

Observable 3: Hypersonic Velocities Without Signatures

This simply means that they can go extremely fast, like the TicTac that dropped 80,000 feet in less than a second—or 30 times the speed of sound—without a sonic boom. That shouldn't be possible.

Observable 4: Low-Observability & Cloaking

Even when these objects are observed, getting a clear and detailed view of them—either through pilot sightings, radar or other means—remains difficult.

Observable 5: Transmedium Travel

Basically, these craft are able to transition seamlessly between flying in completely different environments. They don't just fly through the air, but through the waters of our oceans and through the vacuum of space.

A New Paradigm Emerges

I mean—*what?* This is the guy who used to run the government program that studied this stuff and this is what he is saying. What has he seen that makes him say this? What else don't we know? Are these videos one-off anomalies or just the very tip of an unfathomable iceberg?

I don't know. I seriously have no idea. I've spent a weird amount of time on this subject over the last several months, and the deeper I dig the more questions I have.

I've always considered myself to be an extremely logical person. I grew up in a family of doctors, nurses, and educators where science and critical thinking weren't just valued in an abstract way, but were an active and integral part of my upbringing. While most kids were watching *Barney & Friends*, my dad would teach my brother and I everything from what he was learning in medical school to the physics of falling objects. My mom would regularly bribe us with treats and shower us with praise when we could solve the critical thinking exercises that she prepared for her college-age students. Having a lifelong love of both treats and praise, I got pretty good at it.

While our upbringing ensured that we were always the weirdest kids at any of the many private schools we cycled through as my parents moved us around to pursue their degrees—it also taught me how to think. I've always been grateful for that—and if I'm being honest, it's something that I've taken a lot of pride in.

So when the government came out and confirmed that these videos that had been floating around since 2017 weren't just authentic, but that they had no idea what the craft captured in them actually were, it shook me to my core. Because within that announcement was the realization that I couldn't trust myself—I couldn't trust my mind to sort out fantasy from reality.

The realization didn't hit me like a thunderbolt. It was more like an itchy tag in a new sweater, or a hangnail, or an inflamed taste bud—just this low-level agitation that slowly grew in the background until suddenly it was all I could focus on.

For the first time in my life I started to truly question, not just my own perception, but the very nature of my reality. Because if those videos are real, it stands to reason that at least *some* of the other videos out there are real. Even if that number is super low—let's say 5%—it's hard to deny that the evidence for the reality of UAPs is anything but overwhelming. And even though I had been a witness myself, I'd dismissed it as "tin foil hat" foolishness without ever giving it a second look.

So that's where it started for me: with the realization that I'd gotten this one dead wrong. And with a desire—fueled by ego as much as curiosity, if I'm being honest—to figure out what I'd missed, and in so doing, reorder my universe back into something orderly and recognizable.

I decided that the best way to do this was to start over. I did my best to wipe any preconceived notions from my head and just look at the body of evidence. As it was suddenly very clear to me that the stigma around the topic of UFOs was a major reason that I dismissed key evidence in the past, I decided to take in information from as many different sources as possible. I would let that information wash over me, I'd take on all of the different ideas and theories without prejudice, weighing them and trying them on for size to see if they fit, and in time, I was certain that patterns and eventually a coherent narrative would arise.

And I gotta tell you—I was wrong again. Because the further I go down this rabbit hole the more I realize that I don't know shit. Not just about UFOs, but about everything. As soon as I began to pull on this one thread, everything I thought I knew began to unravel. The real story of the human race and this phenomenon—whatever it is—touches everything from ancient civilizations to consciousness to the very existence and nature of God. It's called literally every single belief I've ever had into question. And I've got to be honest with you: although I have a dazzling number and array of data points, I have no idea what the fuck any of this actually means.

So that's why I'm doing this. That's why I'm talking to you right now. Because I want to take you on this journey down the UFO rabbit hole with me. It's partly a selfless endeavor, because although these questions I've been asking have yet to find any kind of real answers, the process of asking has made me a better person, a more open person, a more humble person, a more spiritual person, a more hopeful person—and I'd like to share that.

But also, my motivations are partly selfish because I need more people to talk to about this. Because the thing you'll find very quickly is that this path is lonely.

We've been trained and conditioned our whole lives to dismiss, ridicule, and mock anyone who does so much as dare to approach this topic with any degree of seriousness. They're immediately labeled as crazy. They're kooks. And when you start bringing this topic up in polite society you see how immediate and reflexive this response is in people, and how hard that mindset is to overcome, even when armed with declassified videos, a Congressional task force, and an unequivocal admission from the U.S. government that this phenomena is, in fact, real.

So I want to talk about all of it—all of the weird twists and turns, all of the wild theories, the colorful and unlikely cast of characters currently at the center of this mystery, and the massive body of evidence that has been hiding in plain sight, maybe since the dawn of time.

Navigating The New Paradigm

But first, I think it's important for us to establish some guard rails. If there is any lesson to be learned from these last few chaotic and unprecedented years it's the very real dangers of misinformation. When you let go of the rope of the established narrative, and find that there is nothing immediately solid to hold onto, it's easy to spin off into space—and before you know it you're holding up a pizza joint because you're convinced it's a front for Hilary Clinton's child trafficking ring. Or you're up all night on QAnon message boards waiting for messages about Donald Trump's secret war against an elite cabal of child-eating satan worshipers.

Yikes. Yikes on bikes. That will not be us, OK? Not on my watch. And to make sure that we don't lose anyone to the dark side on this journey, we need to first equip ourselves with the tools to make sure we don't lose ourselves out there.

Tool #1: The recognition and belief that people are mostly good

Here's the thing: a certain amount of conspiracy theories actually are true. That's not an opinion. That's a fact. Our own government, and governments around the world, regularly declassify documents from decades past that prove in no uncertain terms that false flags, misinformation campaigns, secret black budget projects, secret coups—these things are all real and are carried out with relative frequency and on a breathtaking scale. So although we've all been trained to reject conspiracy theories outright, the reality is that the only idea more ridiculous and fundamentally dangerous than a conspiracy theory is the absurd and nonsensical belief that there are no conspiracies.

But that realization brings us to an equally dangerous place. Because if some conspiracy theories are true—if our government and the powers that be are truly capable of pulling off such elaborate, secretive campaigns on a global scale—then maybe they are *all* true. Maybe *everything* is a lie. Then suddenly every dubious, shadowy connection becomes evidence of an elaborate plot. And in that place you no longer have anything to anchor you, there are no cardinal directions, there is no up or down. You have truly gone through the looking glass, and you're just lost. Many people who go to that place in their minds never make it back.

To me, the only way out of that hole is to hold tightly to the belief that people are mostly good. You have to believe that most people, when presented with a moral or ethical quandary—particularly one with high stakes for the whole of humanity—will strive to do the right thing. And you have to believe that, although they have the same personal flaws and foibles as the rest of us, that most of the people who work in our government and in our armed services are truly there with the intention to serve and work toward the best interests of their fellow citizens.

I know that for many of you reading right now that idea is more ludicrous and hard to swallow than anything I've said about UFOs. And unfortunately, if you don't believe that people are mostly good, I don't think that there is any amount of evidence that I could present to you to make you believe otherwise. But if you do find yourself bristling at this idea, I hope you'll just let this one marinate for a little bit and if you're feeling brave, maybe start to challenge your assumptions in the lab of your own life.

And it's that last part that is the most important part—you have to examine this idea within the confines of your own lived experience, *not* what you see on the internet or on TV. The unfortunate reality of our current stage of technological evolution is that all day long you're being bombarded with images and messages that are meant to manipulate you to take action—to click, to buy, to read more, to follow, to subscribe, to vote, to believe.

And you don't get that kind of action out of people with subtlety and nuance. You don't get it with reassurance and a measured approach. You get it by speaking to people's most basic and primal instincts. You get it by creating fear and uncertainty. You get it by breeding suspicion and hate. Everything is a problem to be solved. Everything is a crisis. And with each new nightmare that unfolds there are a million different solutions—that you can have for a price.

So if you are really going to examine the idea that people are mostly good, you have to look at your actual life. My guess is that if you know any mustache-twisting villains that they are few and far between. What you probably have instead is a bunch of flawed but essentially decent people who are doing their best—and who sometimes can really surprise you by doing something completely selfless. They may not be perfect, but they are mostly good.

Tool #2: The recognition that correlation does not imply causation

Correlation refers to a mutual relationship or connection between two or more things. And causation means, obviously, to cause something. So basically, just because two (or more) things happen at the same time or in the same place or seem in some other way to be connected to each other, that isn't enough to prove that one thing *caused* the other. In fact, they may not actually be related in any real way at all.

A perfect recent example of this is the belief that 5G causes COVID-19. This was a conspiracy theory that caught fire pretty early on during the pandemic when people started to point out that if you looked at a heat map of the places where 5G had been rolled out and a heat map of where the biggest outbreaks of the virus were occurring, you were essentially looking at the same map, causing many people to jump to the conclusion that one was causing the other.

However, knowing what we do about how COVID spreads, it makes perfect sense that big cities and more densely populated areas bore the brunt of the initial outbreaks. And where did 5G roll out first? In big cities and densely populated areas. So while those heat maps could give the impression that these two things were related in some way, they actually had nothing to do with each other at all.

When you let go of the rope of the established narrative and start seeing connections between things that you never saw before, it's a rush. It's honestly exhilarating. Although I'm being entirely sincere when I say that I have no idea what is actually going on, the things that I've discovered while going down this rabbit hole have, in some ways, made the world make sense to me in a way that it never has before. Things that once seemed random or meaningless are suddenly imbued with a deep meaning and seem to make perfect sense inside of the great order of things. And it can be very, very tempting to chase that feeling—*to misidentify it as proof that you're right*—and in doing so to begin to see all kinds of connections between things that aren't really there.

So on this journey it's important that we notice correlations because they can *sometimes* be a roadmap that points us to our next clue. But they can also be nothing at all. So we need to reserve judgment on assuming real connection or causation until we can prove categorically that those connections really exist.

Tool #3: The recognition that the truth is a destination that you can approach forever and never arrive

Renowned physicists, Leopold Infeld and Albert Einstein. wrote the following in their seminal 1938 work *The Evolution of Physics*:

"In our endeavor to understand reality we are somewhat like a man trying to understand the mechanism of a closed watch. He sees the face and the moving hands, even hears its ticking, but he has no way of opening the case. If he is ingenious he may form some picture of a mechanism which could be responsible for all of the things he observes, but he may never be quite sure his picture is the only one which could explain his observations. He will never be able to compare his picture with the real mechanism and he cannot even imagine the possibility of the meaning of such a comparison."

I love this passage so much because it takes the pressure off of trying to find out the "Truth" with a capital "T". Even Einstein, one of the most brilliant minds of his or any time, recognized that his attempts to describe the true nature of reality were ultimately entirely futile—so what hope do any of us have?

The best we could ever hope to achieve is to create theoretical models that more or less explain the world as we experience and observe it. And those theories that most closely describe and predict what we experience and observe are what we call the laws of science. However, we have no way of knowing in our earthly forms—or perhaps ever—what the truth behind our reality really is. To know that would be to know the mind of God.

So it's important that we maintain our humility as we go on this journey. The truth is something that we can pursue together, but no matter how far we go, we'll never get there entirely. And that's OK. It can even be beautiful if you look at it in a certain light.

And more importantly, that humility is a reminder that none of this is worth fighting over and we're all on the same side. As you'll see as we move forward, there are countless different theories in the UFO community about what these things really are, about who or what is behind it all, and about who the good guys and the bad guys are—and the warring and infighting among different factions can get pretty intense. Accusations are constantly being thrown. People can be heroes one moment and then be called "conmen" and "shills" the next. It's a mess.

But as long as we follow this third rule, we can fly above all of that. We don't know the truth. It's likely that no one does. Any consistent and coherent line of questioning about UFOs will inevitably lead you to deeper questions about the very nature of reality and the meaning of our existence— and on those topics, none of us are any better than Einstein. Even people with deeply held beliefs about God and religion have to admit that it's a matter of faith. Because we simply don't know. And when you look at it that way, what possible reason could there be to fight about it?

So, now that we've gotten our tools together to hold ourselves accountable on this journey, I feel it's also important that, as your guide, I be accountable to all of you.

Listen—above all else, I want this journey to be fun. I'm doing this because exploring all of this is really fun for me and I want to share it with you. But I also think it's important to tell you that I approach this topic with a lot of respect, reverence—and, I'm not going to lie, a healthy amount of fear. And that fear isn't even necessarily about the phenomena itself (although there are aspects of it that honestly do terrify me), but rather just an awareness that dealing with this kind of information can be dangerous in very real and insidious ways. In a post-QAnon world, I can't deny the risks. I don't want to be responsible for leading anyone astray or down a rabbit hole that they can't climb out of.

So what I ask of you through this process is that as I start to tell you about fantastical things, that you remain skeptical. Check my work. Dig deeper. Make up your own mind. And my commitment to you is that I will make it as easy as possible for you to do so.

Promise #1: I will show my work.

I don't expect you to take my word on any of this. If I make a claim to you, I should be able to back it up. Showing my work in this way is important because when it comes to this particular area of research, vetting the credibility of sources is complicated to say the least. There are a few reasons for that:

The first is that, as we'll discuss in an upcoming chapter, the mechanisms of our press and academic institutions have not historically been friendly or open to ideas that fall outside of the established narrative. The 2017 *New York Times* article was so shocking, not just because of its contents, but because the mainstream media has so rarely been willing to approach this topic with anything other than a sort of snickering voyeurism—a lighted-hearted moment when anchors can chuckle and exchange knowing looks before sending you over to Bob with the weather. Academia has been similarly avoidant and, despite what the *Indiana Jones* franchise would have us believe, research dollars for anything that could be potentially classified as "paranormal" are almost non-existent.

Because of this, even the sources that we assume to be the most credible in normal circumstances can fall into question. We can't just take them at face value, but rather, we have to put those sources into the context of the time, the identity and motivations of the author, and other clues. And when those other clues include things like heavily redacted government documents, blurry or potentially altered photos and videos, eyewitness testimony, and even ancient texts, it can be nearly impossible to determine which pieces of evidence we can trust.

And frankly, I don't think that you should let me (or anyone else) determine that for you. When there is this level of ambiguity, you have to make up your own mind. By providing you with all of my source materials (see the Bibliography at the end of this book), you'll have the opportunity to do just that.

I will be accountable.

I won't just show my work, but I promise to remain open minded to criticism and feedback. If I make a mistake or am presented with new evidence that substantially changes the facts as I've presented them to you, I will own up to that and will publicly correct my work.

And, for the record, I fully expect this to happen at some point. This is an enormously complex field of study with new developments and revelations coming out almost daily at this point. I'm only one person, and despite my best efforts, I'm bound to get it wrong at some point. And when I do, I'll correct myself.

I will do my best to present a balanced narrative, and to be honest about my biases.

My goal here isn't to spin any particular narrative, but rather to lay out as many of the varied theories and perspectives as I can so that you can connect the dots yourself. In doing so, it will always be my intention to be balanced, fair, and honest.

However, the reason that I'm doing all of this is that I am a giant nerd who is absolutely obsessed with this topic. My own biases are sure to come through, and I think that's OK—as long as I'm honest with you about what they are. I have favorite theories and favorite characters in all of this. I have opinions and I'm highly excitable. I can't do any of this in an authentic way without all of that coming through. However, I'll be transparent with you, and I'll always present you with the alternative view.

It's also important to mention that there is one theory that I will touch on, but that I probably won't give nearly as much time as its alternatives. That theory is the idea that the government is lying about UFOs and that the recent move toward disclosure is a calculated misinformation campaign designed to cover up the fact that these craft are actually ours. Or perhaps it's the prelude to a pretend alien invasion intended to engender support for unprecedented levels of black budget defense spending.

For reasons that we'll get into in future chapters, I don't personally think that's the case. However, that's not why I won't be spending as much time on that theory. The main reason is that it's just...depressing. And while each of the other theories are sobering and terrifying in their own way, none of them are as darkly cynical as that.

There are two doors before us. Behind Door #1 is all of the mystery and potential of the universe, and behind Door #2 is only oblivion. I strongly believe that what we think and where we choose to put our attention matters in very real and tangible ways, so I choose to spend as little time as possible behind Door #2.

Although, I grudgingly admit that even though I hate that theory and find it to be grotesque—it's a legitimate possibility.

I won't try to convince you of anything with regards to this phenomenon.

I think it's important for us all to go into this with our eyes wide open about the fact that "toxic conspiracy thinking" and a "cult mentality" are very near neighbors—and in flirting with one, we are inevitably going to be dangerously close to the other. That's why I not only strongly believe that it's not my place to tell you what to believe about any of this, but I'd caution you strongly against putting too much stock in anyone who tries to.

The only thing that I am interested in convincing you of through this journey—and I sincerely hope that I am successful in doing so—is that the universe is infinitely more vast, complex, and full of beauty than we ever imagined, and that the potential of humanity, both as a species and as individuals, is as profound as it is limitless.

My name is Kelly Chase, by the way. It's a pleasure to meet you, friend. Welcome to the UFO rabbit hole.

2 Are UFOs Human Technology?

If you're anything like me, the only thing you want to talk about right now is what these things could possibly be. So let's just dive into it.

The first and most obvious possibility is that the things that we're seeing in the sky are ours—as in, they are heretofore unknown, terrestrial, man-made technology that has somehow been kept a secret from the general public.

So let's entertain this idea. If these are man-made objects displaying technology that is advanced far beyond anything that we've ever seen before, the immediate question becomes, "OK, but who exactly is making them?

And considering the insane amount of capital that would be needed to develop something so far ahead of our current understanding, there can really only be two culprits—the government (whether ours or someone else's) or private aerospace and defense companies.

It's worth noting that there are some people out there who think it could be some rogue Tony Stark-style billionaire genius developing this on their own—and for some people, this person is specifically Elon Musk. And all I'll say to that is that you shouldn't listen to people who think that Elon Musk is a real-life Tony Stark and not a real-life Philip K. Dick villain.

Also, consider that billionaires Sir. Richard Branson, Jeff Besos, and Elon Musk have all created aerospace companies that are tripping over themselves to secure lucrative government contracts for tasks as mundane as shooting satellites into space. Why would they be wasting their time with that if they'd already developed paradigm-breaking technology? It doesn't make a ton of sense.

It's also unlikely that a private company would be able to develop this kind of technology without the government catching on. And even if they were able to do so, a company's sole purpose is to generate profits for shareholders. So even if they did have this kind of technology, their first move would be to turn around and try to sell it to the only people who would have the cash to buy it—the governments of world superpowers. So we can pretty safely rule out private companies.

However, while it's virtually impossible that any private individual or enterprise would be able to develop this kind of technology without the government knowing about it, it's *also* unlikely that the government would be able to develop this kind of technology—and keep it quiet—without the assistance of private entities.

The U.S. government has a long history of outsourcing top-secret projects to private companies, allowing those projects to dodge the oversight they'd be subjected to if they were housed within a government agency. For example, Lockheed Martin's notoriously secretive Skunk Works division was started in 1943, in the heat of WWII, when the U.S. government needed to quickly develop the country's first jet fighter to compete with the new German jets that were appearing in the skies over Europe.

Four years later, they developed the U-2, the very first dedicated spy plane. While the C.I.A. originally tried to pass it off as a high-altitude plane developed for weather research, it was actually used to take crystal clear photos of the Soviet Union from 70,000 feet in the air.

So there is a lot of precedence for the U.S. government and private aerospace and defense companies working on projects in this sort of hand-in-glove manner. And if these craft are, in fact, the property of the U.S. government, they were almost certainly developed in partnership with private companies.

So that's one possibility.

The other possibility is even more disturbing. Could they belong to one of our adversaries? Could Russia or China have developed this technology without us knowing about it? It's technically feasible, but at this point it seems unlikely.

First of all, it would represent the largest intelligence failure in U.S. history, and given the scale and power of the United States intelligence apparatus that would be a stunning development, to say the least.

The other issue with this theory is that the earliest of the declassified Navy videos we discussed in the last chapter is from the Nimitz incident in 2004. So we have confirmation from the Pentagon that this phenomenon has been going on *at least* since then. If our adversaries had this kind of technology almost 20 years ago, it begs the question—*what are they waiting for?*

Looking back at the five observables outlined by former AATIP Director Lue Elizondo—anti-gravity, instantaneous acceleration, cloaking, hypersonic speed without signatures, and trans-medium travel—any one of those things would represent an absolute paradigm shift not just in weapons, defense, and intelligence systems, but in the ways in which we produce and use energy. The potential and the possibilities are sweeping and profound.

Meanwhile, we've spent the last 20 years engaged in both covert and overt wars in the Middle East and multiple squabbles and proxy wars with Russia, China and others around the globe. The level of technology displayed by these objects is more than sufficient to bring the U.S. military to heel and seize control of the global structures of power. If one of our adversaries was holding that kind of trump card, why wouldn't they play it?

It's also hard to think of a logical reason why the Pentagon would admit to the existence of these craft—and admit that they have no idea what they are—if they thought that there was even a sliver of a possibility that they could belong to another country. That is just not the kind of intelligence that they would ever offer up. It would be dangerous to let the enemy know that they'd caught you flat-footed, and that you didn't know what you were dealing with. You'd also lose the competitive advantage of other countries assuming that the technology could possibly be American.

We only need to look at the Cold War to understand how a government would approach the threat of another country potentially developing weapons and technology that are far superior to their own. I'm not saying that it's not possible that these craft could belong to another country, but if that's the case this isn't at all how you'd expect the U.S. government to respond. Do with that what you will.

Evidence That UFOs Could Be Human Technology

So let's assume for a minute that the Pentagon *is* lying for some reason, and they do know what these craft are because they belong to us. What evidence is there that could support this theory?

The Government Has Lied To Us About UFOs Before

Most people don't realize how much of our perception of the UFO phenomenon is shaped by some pretty seriously revisionist history. Having been raised in a culture and an environment where talk of UFOs and aliens has been consistently responded to with ridicule and mockery, it can be hard to imagine that there was a time when this topic was taken seriously by the government, scientists, and the public-at-large—but there was.

You've probably heard about the Roswell incident in 1947 where witnesses claim to have seen a vehicle that can only be described as a flying saucer that had allegedly crash-landed at a ranch in New Mexico. That very day, the U.S. Intelligence Office of the 509th Bombardment Group at Roswell Army Airfield confirmed to the local evening newspaper, the *Roswell Daily Record* that they had recovered a "flying disc", before changing their story the next day to say that they had been mistaken and it was actually a crashed weather balloon.

The truth of what actually happened isn't really important to this conversation, so I don't want to get bogged down there. All that matters right now is how the government responded to this incident—and to the escalating rash of UFO sightings that were happening across the country at that time.

Because, while outwardly brushing off these incidents as being easily explainable by known, natural phenomena, internally, the U.S. military was taking this potential threat very seriously.

In September 1947, just two months after the Roswell incident, General Nathan Twining—a former combat fighter pilot, WWII commander, and the head of the United States Air Material Command—wrote in a now famous letter that the UFO phenomenon was "something real and not visionary or fictitious". He described the existence of "metallic" looking discs that showed incredibly advanced capabilities and that appeared to be intelligently piloted or remotely controlled. He subsequently recommended the formation of what became known as Project Sign to study this phenomenon.

Project Sign eventually was renamed as Project Grudge and then as Project Blue Book in 1952. While early reports from Project Sign suggested that these UAPs could potentially be of Soviet origin, by 1952 the multitude of reports that Project Blue Book was studying—many from highly credible witnesses, including those within the US military—were being routinely dismissed as everything from rare atmospheric anomalies to mass hysteria to deliberate hoaxes in their reports. However, it's clear from now declassified documents that while the government was reassuring the public, it was still treating this phenomenon as an active threat.

In July 1952, in a series of incidents over multiple days, at least ten glowing UFOs were seen by countless people in the skies above the White House and U.S. Capitol building—at one point for 6 straight hours—and were tracked on radar at Washington D.C.'s National airport. Fighter jets were scrambled, but each time they got close to the UFOs they disappeared or took off at speeds that made them impossible to chase. The incident was front-page news around the globe and even President Truman demanded answers.

The official explanation, which was delivered to the American people by U.S. Air Force Director of Intelligence, Major General John Samford, in the biggest press conference since WWII, dismissed these incredible events as a freak "weather phenomena". Nothing to see here. And yet just three months later the C.I.A.'s Assistant Director of Scientific Intelligence wrote the following in a now declassified secret memo:

"'Flying saucers pose two elements of danger which have national security implications. The first involves mass psychological considerations and the second concerns the vulnerability of the United States to air attack."

Whether these UFO incidents being reported were legitimate or not doesn't change the only conclusion that can really be drawn here, which is that the official story that was being fed to the public was not at all an accurate representation of what was happening behind the scenes. And considering that, as was first exposed in the 2017 article in *The New York Times,* programs for studying UAPs have continued to exist within the Pentagon up through the present day, it's nearly impossible to accept the idea that the government thought this was all just pranksters and swamp gas.

So we do know *for a fact* that the government has a record of being dishonest about what it knows and doesn't know about UFOs. So is there any evidence that they are lying now when they say that they don't know what this phenomenon is? Is there any evidence to suggest that this could actually be our technology?

It turns out that there is. It's extremely circumstantial, but it's compelling enough to be noted

The Missing $21 Trillion

As I mentioned a little earlier, one of the most obvious things about developing this sort of super-advanced technology is that it would be expensive. Like really, really expensive. We're talking about the kinds of numbers that just break your brain and make your head spin with their sheer magnitude.

Let me put it this way. The most expensive military plane in the world is the Northrop Grumman B-2 Spirit. It costs $2.1 billion and $135,000/hour to operate—and that's with technology that we understand. The cost to develop the kind of ridiculously advanced craft that are being tracked in our skies and beneath our oceans would be absolutely astronomical—potentially into the hundreds of billions or even trillions of dollars.

The spending power of the United States is mind-boggling, but could the government really spend and hide trillions of dollars without us knowing?

It turns out that they actually do this all the time through unsupported budgetary adjustments. The adjustments are done to balance the books when there is an expense that can't be accounted for with the proper documentation. Basically, the money was spent, but no one can find the receipt.

Except we aren't talking about minor discrepancies here. In 2015 alone, the U.S. Army failed to provide adequate support for $6.5 trillion in spending. And between 1998 and 2015, it's estimated that $21 trillion in spending from the Department of Defense and the Department of Housing and Urban Development is completely unaccounted for. The money was spent on something, we just have no idea what.

Now it's important to remember that, as we discussed in the last chapter, correlation does not equal causation. So just because trillions of dollars disappeared doesn't mean that the U.S. spent it on developing the paradigm-smashing technological objects that are being recorded in our skies. This is just about assessing what is possible so that we can eliminate the impossible.

So we've established that the government has a track record of lying to the American people about this phenomenon. And with $21 trillion dollars of taxpayer money seemingly "in the wind", they could conceivably have had the money to fund a massive black budget project like this.

Could The Whole Thing Be A Lie?

And there's another possibility as well, which is that the Pentagon is actually lying about all of this, and the technology doesn't exist at all. They faked the videos and the slow crawl to disclosure that the UFO community has been watching with rapt attention is nothing more than an elaborately orchestrated psyop designed to manufacture a fake "alien threat" in order to...I don't know. Something about money and the military-industrial complex and endless secret wars.

I told you. I fucking hate this theory. It's not that I don't admit that it could have some merit, but I can't think of anything more bleak or depressing.

The good news is that, as you'll see as we dig into the history of the United States with regard to this phenomenon, it's at least my personal opinion that this has all been way too well documented for way too long for this to be simply a sucky, stupid, cynical, psyop. So there is hope for us yet.

But the biggest and most obvious evidence that this technology could be human is that—although "the government is faking an alien threat in order to consolidate power and get a rubber stamp for military spending" sounds pretty crazy when you say it aloud—as you'll see, it's the least crazy theory, by far. Don't get me wrong, if that's what's actually going on it's one of the wildest things that's ever happened in human history, but unlike the other options, it doesn't require us to go back to the drawing board on literally everything we know about history, science, and our place in the universe.

And based on that fact alone, you might be thinking—OK, well then why are we even considering any other options? If this is the only explanation that makes sense given our current understanding, then why are we wasting our time on any of this other nonsense?

I hear you on that. I really do. And that was basically my stance on the issue when I first started down this rabbit hole. After all, I come from a medical family and one of the first things that they teach new doctors about diagnosing illnesses is that "when you hear hoofbeats, think horses—not zebras." The extraordinary does happen, but very, very rarely. That's what makes it extraordinary. And to justify shredding everything we know about the universe and our place in it, we'll need evidence of something that is more than just extraordinary.

Shifting Paradigms

But let's not forget that there have been major paradigm shifts throughout human history that have turned everything on its head—revelations that forced us to go back to the drawing board as a species on all that we "know" to be true.

The Heliocentric Solar System

For example, dating back to the time of Aristotle in the 4th century B.C. the prevailing belief was that the cosmos (which at the time consisted of everything we could see in the sky with our naked eye) revolved around the Earth in perfectly circular orbits. The Earth was quite literally the center of God's creation.

It wasn't until almost 600 years later, in the second century A.D., that the astronomer Ptolemy recognized through observation that the heavenly bodies didn't actually appear to move in circular orbits. They moved forward and backward across the night sky in ways that shouldn't have been possible if their orbits were truly circular.

This caused Ptolemy to develop his own model that showed the celestial bodies in more elaborate orbits that accounted for their movement in the sky well enough that it became the primary predictive model of astronomy for the next thousand years. But he was still working with the same fundamental— and false—assumptions. He assumed that the planets, moon, and sun were orbiting the earth in perfectly circular orbits—and that it was only a matter of perspective that made it appear otherwise.

So Ptolemy was successful in developing a model of the cosmos that was accurate enough to be predictive, but the truth of what he was observing was still hidden from him and the rest of humanity for the next 1400 years. Now there's a humbling thought.

It wasn't until 1543, that Nicolaus Copernicus detailed his radical theory of the Universe in which the Earth, along with the other planets, rotated around the Sun. And even then, his theory took more than a century—and the invention of the telescope—to become widely accepted.

And even then, there were still pieces of the puzzle missing. But those pieces of the puzzle were so revolutionary and so far beyond our understanding at the time that hardly anyone even really noticed.

Gravity & Relativity

Once Issac Newton defined the laws of motion and gravity in his book *Mathematical Principles of Natural Philosophy* almost 150 years later in 1687, scientists believed that they finally had it all figured out. And after 210 years of feeling smug, in 1897, the physicist William Thomson, Lord Kelvin famously concluded: "There is nothing new to be discovered in physics now. All that remains is more and more precise measurement."

Which is awkward for him considering that less than ten years later Albert Einstein began to publish work on his theory of relativity. In what amounts to one of the greatest and most profound scientific discoveries in human history, Einstein found that Newtonian physics depended on the assumption that mass, time, and distance are constant regardless of where you measure them.

The theory of relativity treats time, space, and mass as fluid things, defined by an observer's position and frame of reference. All of us on the Earth are in a single frame of reference, but an astronaut in a fast-moving spaceship would be in a different frame of reference.

If you're measuring something from a single frame of reference, the laws of classical physics, including Newton's laws, hold true. But Newton's laws can't explain the differences in motion, mass, distance, and time that result when objects are observed from two very different frames of reference.

For example, a clock on a satellite orbiting the Earth at 14,000 km/hr in an orbit that circles the Earth twice per day is moving much faster than any of the clocks on Earth. Because of the theory of relativity we actually know that clocks that are moving faster tick slower. And if this clock is 20,000 km above the Earth, it's experiencing one quarter of the amount of gravity as clocks on Earth, which will actually make it tick faster. The net impact would be that the clock on the satellite would move 38 microseconds per day faster than the clocks on the ground.

If that hurts your brain a little, you're not alone. It's a super advanced concept that requires that you put aside everything that you think you know about the fabric of your reality. You can't really do anything but marvel at the kind of brain that could look so far beyond its own observed experience to understand something so profound.

But even Einstein could be a victim of his own assumptions and prejudices. Einstein actually spent most of his life thinking that there was a flaw in his theory of relativity. His math seemed to suggest that the Universe was expanding—and doing so at an increasing rate. But the idea was so outrageous at the time that Einstein dismissed it, and instead introduced the idea of a "cosmological constant" to counter what the math was showing. It was essentially a way to fudge the numbers to make his theory conform to what he believed *must* be true.

But that all changed when Edwin Hubble proved that the Universe was, in fact, expanding at an increasing rate. Einstein called his personal refusal to accept what the math was showing him because he didn't think it was possible "the greatest blunder" of his career.

So where I'm going with all of this is that, although they are rare, throughout history there has been a progression of massive paradigm shifts in the way that we view ourselves and our place in the cosmos. And right up until those discoveries were made—and often until long after—people were convinced that the science of their day had already answered all the questions that could be asked.

We shouldn't allow ourselves to recognize the arrogance of that without also turning inward and recognizing that arrogance within ourselves. It's easy to look around at the world, and at our stunning rate of technological advancement, and believe as Lord Kelvin did that there is nothing new to be discovered, only more precise measurement.

Unfortunately for our egos, that's almost certainly not true. Because while the Theory of Relativity briefly unified physics into a mostly coherent whole, quantum physics blew it apart again—and we still haven't quite figured out how to clean up that mess

Quantum Mechanics

You've probably heard about the famous "double-slit" experiment, but here's a quick refresher in case you don't sit around thinking about quantum mechanics all day:

OK, so imagine that you have a board with two vertical slits cut into it. If you shine a light at the board, the light waves will go through the slits and as the waves of light overlap and interfere with each other, it will create a repeating pattern on the wall behind the board called an interference pattern.

But as you probably recall, light can also exist as a particle called a photon. And what's crazy is that when scientists fire a single photon at a time at the two slits, it still creates an interference pattern on the wall, indicating that each photon is somehow going through both slits, meaning that it is behaving like a wave and not like a single particle.

Even weirder is that if scientists then introduce a sensor to observe each photon going through the slits, then the photon will then act like a photon and go through only one slit. And as they continue to fire one photon at a time, the photons end up in two straight lines instead of in the interference pattern created by light waves.

So what this means is that light can behave like a particle and a wave *at the same time* and the *mere act of observing the experiment* will force the light to behave as one or the other.

Putting to the side for the time being that this suggests that consciousness has an active role in the creation of our lived reality (insert mind-blown emoji here), this created a major divide in physics that still hasn't been totally resolved. I'll tell you about it real quick, because who doesn't love a good old fashioned physics beef?

So taking us back to high school physics again real quick, do you remember the Schrödinger's Cat Experiment? For any cat lovers out there, of which I am one, I will preface this by saying that it was a thought experiment, and not something that Schrödinger actually did.

So in this thought experiment, Schrödinger imagined putting a cat in a box with a sealed vial of cyanide, and with a small hammer hanging over the vial. The hammer would be connected to a Geiger counter, which detects radioactivity, and that counter would be near a tiny lump of slightly radioactive metal. The second that the metal released even a tiny bit of radiation, the hammer would smash the vial and the cat would die. Schrödinger's experiment would then basically involve putting the cat into this contraption for a set period of time and then opening the box to find out the cat's fate.

According to quantum mechanics, because the radiation from the lump of metal is composed of subatomic particles, without an observer it both will and will not be emitted while the box is closed, implying that *until it is opened the cat is both dead and alive at the same time.*

And Schrödinger saw that there was a major issue with this. Because it's one thing to say that a photon can go down two different paths at once, and another to say that a cat can be both alive and dead at the same time. It seemed to him that although quantum physics could explain how very tiny objects behave—so well that we've developed lasers, LED lights, and space probes—it was completely at odds with the rest of our observable experience.

However, most of Schrödinger's contemporaries completely dismissed his concerns. Many had no issue with the cat being both alive and dead inside of the box until it was forced into a state of "aliveness" or "deadness" once it was observed—despite having no way to explain how that could be possible.

But most of them, including revered physicist Niels Bohr, thought that the question was meaningless. Bohrs argued in the famous Bohrs-Einstein debates that the inside of the box was, by definition, unobservable—and that only things that can be observed and measured have meaning. So, in other words, this is the quantum equivalent of asking, "If a tree falls in the forest does it make it sound?" And the official answer is, "Don't worry about it."

I don't know about you, but I don't find "don't worry about it" to be a compelling or convincing answer to what is undoubtedly one of the biggest questions in modern physics. But that's exactly where—with a few notable exceptions—most of the scientific community stands on this issue today.

It's the very essence of what it is to be human to want to understand where we come from and why we are here. But we need to be careful that our desire to find those answers doesn't cause us to say "don't worry about it" to the evidence that doesn't make sense within our current paradigm.

So to bring us back to where we started this crazy tangent—the question we were trying to answer was whether it even makes sense for us to consider alternative explanations for the UFO phenomenon besides them being secret human technology, despite the fact that that is the only explanation that doesn't require us to rip up our science and history books.

And, for me, the answer is yes. And the reason why is that there are enough pieces of incongruent data that just don't quite fit together. And for questions this big, I don't think we should accept "don't worry about it" as an answer.

However, we still need to proceed cautiously. As Carl Sagan so wisely said, "Extraordinary claims require extraordinary evidence." We can't just go ripping up the text books until we're absolutely certain about what the data is telling us.

And we also have to accept that we might never get the answers that we're looking for. These sorts of quantum leaps in our understanding about the nature of our reality move on their own timeline. Sometimes, like in the last century, a lot happens all at once—but sometimes humanity works on a problem for hundreds of years before collecting enough evidence to make a major breakthrough.

Right now, we're at the point where we have enough information to start to know which questions to ask. But there's no way for us to know how long it might be before these mysteries reveal their secrets to us.

Are you still in? Then let's do this.

What Is The Evidence That UFOs Aren't Human Technology?

First of all, although the declassified Navy videos were all from the 2000s, the UFO phenomenon has been well-documented for *at least* the past 80 years. Before the Pentagon made its announcement that UFOs are real and that they don't know what they are, we had the luxury of brushing off the thousands upon thousands of UFO sightings that have occurred. It was a hoax. It was swamp gas reflecting off Venus. It was a weather balloon. And to be fair, many of those sightings—maybe even the vast majority—have rational explanations.

But now that the government is saying that UFOs exist, we no longer have the luxury of saying "don't worry about it."

Shaking The Stigma

I don't know about you, but when I got to this point, I actually felt kind of silly. With so much evidence that something weird was going on in the sky, how did I dismiss it so thoroughly and without question?

Because although the government has done a masterful job of not just denying that the phenomenon exists, but making it taboo for people to even talk about it, I'm shocked at the level of cognitive dissonance that it took to not see something that seems so obvious now.

We've been conditioned into believing carefully crafted stereotypes about people who report encountering UFOs. We assume they're uneducated, ignorant, and they almost certainly have a screw loose. But there are scores of cases where the people making these reports are highly credible people who have fled from the spotlight, never made a dime off their stories, and have had their lives and careers destroyed by coming forward—leading one to ask, why would they lie?

And there are other cases, where the credibility of the witnesses themselves becomes almost irrelevant due to the sheer number of people who saw the same thing at the same time. For example, in March 1997, thousands of people reported seeing an enormous v-shaped craft with five lights on the bottom flying low over Phoenix, Arizona—with multiple people recording it from vantage points across the city.

Witnesses have never been satisfied with the government's explanation that the lights were caused by flares dropped by the Air Force during training exercises, and continue to search for answers. Even the former governor who played a major role in minimizing and ridiculing the incident in the press has since come forward to apologize to the witnesses. He now says that he saw the craft, as well, and that he could only describe it as "otherworldly".

The UFO Phenomenon Isn't New

The thing that makes this whole thing even stickier is that, although it's generally accepted that the first report of a flying saucer was by a pilot named Kenneth Arnold in 1947 who made his sighting while searching for a lost airplane near Mount Rainier, strange things have been seen in the sky for a long time before that. And though pre-industrial, pre-flight people may not have had the same words and frame of reference as their 20th century counterparts, what they described seeing in the air sounds very similar to modern day UFO sightings.

In his groundbreaking 1969 book, *Passport To Magonia*, venerated ufologist Jacques Vallée presented 923 eyewitness accounts of strange objects and aerial phenomena from around the globe spanning the 100 years between 1868 and 1968. Here are a few:

In July 1868 in Copiago, Chile "a strange 'aerial construction' bearing lights and making engine noises flew low over this town. Local people also described it as a giant bird covered with large scales producing a metallic noise."

In 1877 in Great Britain it was reported that, "A strange being dressed in tight-fitting clothes and a shining helmet soared over the heads of two sentries who fired without result."

In April 1897 in Everest, Kansas "the whole town saw an object fly under the cloud ceiling. It came down slowly then flew away very fast to the southeast. When directly over the town it swept the ground with its powerful light. It was seen to rise up at fantastic speed until barely discernible, then to come down again and sweep low over the witnesses. At one point it remained stationary for five minutes at the edge of a low cloud, which it illuminated. All could clearly see the silhouette of the craft."

And these are just three of over 900 recorded sightings. Many of them, like these three accounts, were reported years, if not decades, before the Wright brothers first achieved human flight in 1903.

The uncomfortable reality is that strange things have been reported in the sky for centuries and perhaps even millennia—even appearing in the earliest written records and religious texts that we have. It's impossible to know exactly what, if anything, people were actually seeing back then, or how many of those cases were a genuine case of mistaken identity or mass hysteria or a hoax.

And yet, you quickly realize that any effort to draw a line in the sand and declare that any particular incident or time period marks the definitive beginning of the UFO phenomenon would be entirely arbitrary. Whether real or imagined, this is an experience that people have been having for a very long time.

And that's a major problem for the argument that UFOs are human technology. Even if we take the most conservative set of assumptions and say that the 2004 Nimitz incident from the FLIR video was the first ever legitimate UFO encounter, the idea that anyone anywhere on the planet had that technology almost 20 years ago stretches the limits of credulity. And to think that we potentially had it right after WWII when UFOs first started to dominate the public consciousness, seems too far-fetched to even be considered.

And yet, in November of 2021, former director of AATIP, Lue Elizondo, said in an interview with British GQ:

"I have in my possession official U.S. government documentation that describes the exact same vehicle that we now call the Tic Tac [seen by the Nimitz pilots in 2004] being described in the early 1950s and early 1960s and performing in ways that, frankly, can outperform anything we have in our inventory. For some country to have developed hypersonic technology, instantaneous acceleration and basically transmedial travel in the early 1950s is absolutely preposterous."

Because, to be clear, what we are talking about with the technological capabilities shown isn't just a quantum leap forward—it shatters every rule and every paradigm we have.

And even if some of the craft that we're seeing are human technology, it begs the question—where did we get it? Where did it come from? Because if it is us, the timeline of all of this makes it hard to believe that we made these breakthroughs entirely on our own.

And there's also the question of why the government would still be investigating UAPs if we knew what they were. AATIP is not the first government agency to study this phenomenon. This has been studied by multiple government agencies at various levels of secrecy since *at least* December 1947.

So why do all of this if we already know what they are?

Could UFOs Be A Psyop?

For the "everything is a psyop" crowd the most likely answer is that—you guessed it—it's a psyop. They say we can't trust anything that the government, or anyone associated with the government, tells us about UFOs. They point to Lue Elizondo, currently the most prominent face of the disclosure movement, and to his background in counterintelligence and the fact that, despite resigning, he still maintains his clearance as proof that whatever the government is telling us about UAPs is being done to manipulate us to some nefarious end—probably war.

And like I've said, I hate this theory. I hate everything about it. But I'd have to be naive not to consider it. And I have—deeply.

But for me, when I listen to Lue Elizondo speak, I don't hear someone pushing a hateful, fearful agenda. He's not making the case that we should fight these things—in fact, he makes it clear that we're entirely outmatched by this technology to an extent that it makes the idea of fighting with them almost meaningless. How could we fight something that can travel 80,000 feet in under a second?

In fact, Lue actually seems to be saying something deeper and much more profound—something that isn't just about the origin of this phenomenon, but about the origin of our species and the very nature of our reality. In what has become one of his most-quoted answers in an interview, when Lue was asked how he thought people would feel if they knew what he knows, he said "somber".

He later explained his answer further in an interview with Curt Jaimungal on his phenomenal YouTube Channel *Theories of Everything* by saying:

"Imagine everything you've been taught, whether it's through Sunday school or thorough regular formal education in school, what our political leaders have told us. And yes, what our mothers and fathers around the dinner table have told us, or maybe at bedtime about who we are—our background and our past. What if all of that turned out to be not entirely accurate? In fact, the very history of our species, the meaning of what it means to be a human being and our place in this universe—what if all of that is now in question? What if it turns out that a lot of the things we thought were one way, aren't? Are we prepared to have that honest question with ourselves? Are we prepared to recognize that we're not at the top of the food chain, potentially? That we're not the alpha predator? That we are maybe somewhere in the middle?"

And listen to this answer that he gave in an interview on *The UFO Podcast* when asked about research that he was doing with the Lakota people.

"Let's not forget that the First Nations of Saskatchewan—and frankly, most indigenous peoples —have oral traditions that go back, in some cases, thousands and thousands of years. Prehistory from our perspective.

And, very much the same as an F-18 pilot is a trained observer when flying a combat mission, indigenous people are trained observers when it comes to their land. And though it might come from a slightly different paradigm than we're used to in the Western world, indigenous people have a profound sense of history. And it is that history that we're very excited to see if we can see some of the same patterns that we've seen in the current effort regarding AATIP.

But even more importantly than that, I think all indigenous people have lessons that we can learn from. I think we live in a very materialistic world that is heavily dependent upon technology, and indigenous people for millennia have gotten along just fine without it. And their sense of government, their sense of organization, of fairness, is truly profound, and I think if we can show the world just a little bit of that beauty, we might be better off. We might be better off as a civilization. We might be better off as a species.

I'm not going to go into detail, but when I had the distinct honor of going up and meeting the Chief and the Elders of the Bands—the Lakota Bands (they prefer to be called Bands)—it was a profound experience for me. It was an epiphany. I learned so much in the brief time that I had with them. And again, I don't want to share the specifics in this instance, because I'd like to eventually provide that in a much more professional manner for everybody to see.

But it was nothing less than soul-shaking. I mean that sincerely. I realized there's a whole other aspect to humanity that we've forgotten. And indigenous people, despite the challenges that they have faced over the years, whether it's disease or being moved against their will to reservations and the hardships, the discrimination that they've faced over and over again—somehow they've been able to maintain their dignity and their humanity.

For me, it was an understanding of resilience and that there is an indelible part of a human being, whether you can it the 'soul', or the 'Id', or the 'Chi'—whatever 'nam du jur' you want to give it—there's something beyond the biomass of a human body and beyond the electrical synapsis of the brain and intellect that makes each and every one of us distinct.

And I learned that there's great beauty in that. And coming from a hard investigator who's been a 'just the facts, ma'am' kind of guy, there's another aspect to being human here that, not just me, maybe other people have forgotten, as well."

Do you see what I mean?

Honestly, if it wasn't for Lue, I don't know how far down this rabbit hole I would have gone. But the former director of a shadowy government agency whose core mission was to investigate UAPs quit in protest in order to pursue disclosure, and *this* is what he is saying? What could it possibly all mean? I don't know about all of you, but I'm hooked. Wherever this train is going, I'm along for the ride.

And I hope you're coming with me! Because in the next chapter we're going to start exploring some of the alternative explanations for the UFO phenomenon if it is, in fact, non-human technology. And we're going to start with talking about the one word I haven't said yet that you've probably been waiting for—

aliens.

3 Are UFOs Alien Technology?

Now we're really starting to get to the good part. As we discussed in the last two chapters, the Pentagon has confirmed that UFOs are real. However, they've very conspicuously avoided public speculation about what they could be.

What they have said basically boils down to this: there are intelligently controlled technological objects being tracked in our skies and in our oceans that can do things that we can't explain. We don't know what they are exactly, and we don't know who they belong to. But they aren't ours and we don't believe that they belong to any other country.

So what the heck are they? And does this mean what it sounds like it means?

Although the Pentagon has very purposefully not said the words "alien" or "extraterrestrial" they have yet to offer any other explanation. And so the natural conclusion for many is that, if they aren't from here, they must be from "out there".

So what evidence is there that it's aliens? Like everything that has to do with this topic, the answer is more complicated than you might think. So get some snacks, clear your schedule, and buckle up, friends. We've got a lot of ground to cover. And a great place for us to start is with the Fermi paradox.

The Fermi Paradox: Where Is Everybody?

In 1950, Enrico Fermi, a physicist at Los Alamos National Laboratory—who, as a side note, developed the first nuclear reactor—was having lunch with his friends when he wondered aloud, "Where is everybody?"

Fermi had been thinking about the lack of evidence for extraterrestrial life, and it just didn't make sense to him. After all, if the Universe is 14 billion years old and there are over a *billion trillion* stars, how could we possibly be alone? And if we're not alone, shouldn't we have heard from some of our neighbors by now?

Just look at our own development as a species. In less than a million years we went from rudimentary wooden tools to landing on the moon, launching a probe beyond our solar system, and we're starting to eye Mars for colonization. Fermi crunched some numbers and he reasoned that any species with the ability and drive to colonize other worlds would be able to colonize an entire galaxy in about 10 million years.

Now, 10 million years may sound like a crazy long time, but when you think about the fact that our galaxy is 10,000 million years old, you realize that's no time at all.

So given those odds, we almost certainly should have encountered extraterrestrial intelligence by now—but as Fermi asked, "Where is everybody?" And that's the paradox.

There are a bunch of possible explanations for why this is the case, and all of them are super interesting, and also very relevant to this whole "is it aliens?" conversation.

Young Earth, Old Universe

One thing that makes the Fermi paradox particularly perplexing is that the Earth is relatively young in the context of the Universe. The Universe is estimated to be 14 billion years old, but the Earth is only about 4.5 billion years old.

If we think about how difficult it would be for someone from 1000 years ago to comprehend our current level of technology, and then recognize how our technological development has seemed to increase exponentially over the last century, it can be hard to imagine what a civilization that was 1000 years ahead of us would look like—much less one that had a head start of billions of years.

There is this model called the Kardashev Scale that we use to try to wrap our minds around this. The Kardashev Scale is based on the assumption that as a civilization advances and begins to colonize the universe that its population growth and technological developments will require increasing amounts of energy, and it categorizes civilizations by the amount of energy that they are able to harness.

There are 3 main categories: Type I, Type II, and Type III Civilizations.

Type I Civilization

A Type I Civilization is able to harness all of the energy of its planet. Humans actually aren't even a Type I Civilization yet. We're still down at Type 0. And to give you a sense of how far we are from that, right now we're actually only able to produce about 1/100,000th of that energy. But as a Type I Civilization, we'd hypothetically be able to harness all the power of the Earth's oceans, volcanic activity, and more.

Type II Civilization

A Type II Civilization is able to harness all of the energy of its host star. That would take a level of technology that we can scarcely understand. Scientists have proposed that it could be done with a hypothetical megastructure called a Dyson Sphere that would basically be built around the sun to catch every bit of solar radiation.

Type III Civilization

A Type III Civilization is able to harness all of the energy of its galaxy. I'm not even going to pretend to speculate on how that might be done. It truly boggles the mind. However, if a civilization was really billions of years ahead of us in their development, this is the level of advanced civilization that would be possible. And if they're out there, you'd think they'd be super noticeable, especially if there were a lot of them.

So how many of these advanced civilizations do we think might be out there? That's where things get complicated.

The Drake Equation

There is a generally accepted formula for determining how many alien civilizations there should be who have reached a level of technological sophistication that we should be able to detect. This is called the Drake Equation, and while scientists agree on the formula itself, when you start plugging in numbers, there's a lot of debate.

The Drake Equation is:

$$N = R^* \times fp \times ne \times fl \times fi \times fc \times L$$

N   : the number of civilizations in the Milky Way galaxy whose electromagnetic emissions are detectable

R*   : the rate of formation of stars suitable for the development of intelligent life (number per year)

fp   : the fraction of those stars with planetary systems

ne   : the number of planets, per solar system, with an environment suitable for life

fl   : the fraction of suitable planets on which life actually appears

fi   : the fraction of life bearing planets on which intelligent life emerges

fc   : the fraction of civilizations that develop a technology that produces detectable signs of their existence

L   : the average length of time such civilizations produce such signs (years)

Makes a lot of sense, right? The issue with the Drake Equation is that it's one big multiplication problem. So if *even one* of the numbers you plug in is zero—for instance if the number of other planets on which intelligent life can evolve is zero—then the answer is zero and the reason we haven't found anyone else is because they aren't there.

And because that has been our experience, it can be easy to want to stop right here, and say "case closed". The problem is that if we assume that any of those variables in the Drake equation are zero—or close enough to it to effectively be zero—that makes us *very* special. We're not talking about one in a million, but one in one hundred billion — and that's just talking about the planets in our own galaxy.

What are the chances that we are actually that special?

That's a question that is actually impossible to answer, at least with the data we currently have, because of something called the observation selection effect.

Observation Selection Bias

An observation selection effect exists when some property of a thing is correlated with the observer existing in the first place. When such an effect is present the data will be biased, often in significant ways. The more extreme the correlation—basically the more that the thing being observed is related to the observer—the more unusual effects it could have.

So in this case, the thing we're trying to observe is whether or not we are unique on a level that would make finding others like us unlikely. The problem is that our human intellect that we use to even ask that question inherently makes us special. As far as we know, we're the only species in the 4.5 billion year history of this planet that has reached this level of sentient consciousness and intelligence. That's pretty special right there.

So ultimately we really have no way of knowing how special or how not special we are. Even if there was only one intelligent species in the entire Universe, that intelligent species would inevitably ask this question. We could be that one—or we could be one of billions. We simply don't know.

And so that leaves a lot of unanswered questions, but also a lot of possibilities in terms of what could explain the Fermi Paradox. We can't really narrow it down without more information. So let's talk about what those possibilities are.

The Fermi Paradox: All The Possible Explanations

If we get down to basics with the Fermi Paradox, one of two things is true:

1. We haven't found signs of intelligent life outside of our planet because it isn't there. (Which means we're special.)
2. There is intelligent life all over the place, but there's a good explanation for why we haven't found it yet. (Which means we're not special.)

Let's start with the first possibility.

Explanation #1: We haven't found signs of intelligent life outside of our planet because it isn't there.

So this is the world in which we are special, but, as we'll see, that's not necessarily good news for us. Because while Fermi's math calculated that there were potentially thousands of other intelligent civilizations in our galaxy—and other people claim much smaller and much more specific numbers (like 36)—if even one of them is a Type II or a Type III Civilization, their presence should be pretty obvious.

The Great Filter

So if there are supposed to be other intelligent civilizations here and they aren't, then that means that there must be something else going on. And to address this problem, there is the theory of The Great Filter, which is basically the idea that somewhere between the pre-life stage of a planet and the Type III stage, there is a filtering event that all (or very nearly all) species fail to get past.

If that is the case, then our fate is very much tied to where we are in relation to the Great Filter. And there are 3 basic possibilities:

1. We're rare.
2. We're first.

3. We're screwed.

Possibility #1: We're rare.

In this scenario, the Great Filter is already behind us, and we're one of a vanishingly small amount of species that ever make it this far. Yay us!

So if this is the case, then what could the Great Filter have been? It's hard to tell, but there are some strong candidates.

It could be that the conditions that create life are extremely rare, and that the vast majority of planets never even get that far. Or it could be that simple forms of life like single-celled organisms are relatively common, but that it's extremely rare for those to develop into multicellular organisms. Or it could be that complex life like plants and animals as we know them are pretty common, but that developing human-level intelligence is super rare.

It could also be that the Earth itself is super rare. We tend to take for granted how perfectly calibrated our planet is to support life. Just the fact that we have such a big moon that orbits so closely to the planet and is tidally locked with the Earth creates the ocean tides and currents that stabilize our seasons, making our planet particularly hospitable to life.

Maybe it's our moon that has allowed complex life to evolve and flourish long enough to produce intelligent life. It's certainly possible. And actually, we have lots of reasons to believe that our Moon is nothing like other moons. It's an outlier in a lot of ways.

I know I don't have time to talk to you about the Moon right now. We're in the middle of something. But stick a pin in that and we'll get back to it in book two—because the Moon is weird.

Possibility #2: We're first.

If we're not rare, but we still can't find anybody else, could it be that we just got to this stage of development first? Perhaps.

One obvious complication for this possibility though is that, as we discussed, the Earth is relatively young compared to the Universe itself. So how could we be first if we were seemingly late to the party?

One explanation could be that the Universe was just too chaotic for life to develop during the time before the Earth was formed. The Earth may have just popped up at the earliest time where there was enough stability to support the development and evolution of complex life-forms. In this scenario, there wouldn't even necessarily need to be a great filtering event. We just have to be patient until everyone else catches up.

Possibility #3: We're screwed.

The final possibility is that the Great Filter is ahead of us, and we're most likely on a collision course with our ultimate demise as a species. In short—we're screwed.

It could be that once a species reaches a certain level of technological advancement that most—if not all—of them end up destroying themselves. This scenario hits pretty close to home when you think about the absolute havoc that our technological advancements have wreaked on the planet and on each other.

Not only are we destroying the environment, but we currently have 150 times more nuclear bombs on the planet than it would take to effectively end life as we know it. The idea that the Great Filter is ahead of us doesn't feel like that much of a stretch. However, luckily for us, there is still the second possible explanation for the Fermi Paradox.

Explanation #2: There is intelligent life all over the place, but there's a good explanation for why we haven't found it yet.

While the first explanation depends heavily on the idea we're special in some way, this explanation relies on the Mediocrity Principle—which basically assumes that our planet and our solar system are pretty average and not different from most other solar systems in any meaningful way.

So if we're not special in any way, then there should be other intelligent species in the galaxy. And so if we can't see them, we have to assume that there is just something that is preventing us from detecting them. As for what that "something" could be, the possibilities are pretty mind-bending. And I'm not going to lie—some of them are pretty terrifying.

Possibility #1: We're assholes.

It could be that we're just projecting the idea that a highly advanced technological society would want to do things like colonize space and find increasingly elaborate ways to harness more energy to meet perpetually rising demand because those are the kind of asshole moves that we would make.

And yet those are the very assumptions upon which the Kardashev Scale is based. So our idea of what a Type I, Type II, and Type III Civilization would be could just be way off base. Maybe most higher intelligences are smart enough to pursue more worthy and less destructive pursuits. Maybe at a certain point they upload their consciousness to a nebula and just vibe. It's possible that they're out there and we just don't know what we're looking at because we expect them all to be dicks like us.

Possibility #2: There are way bigger assholes out there than us out there.

It could be that there are scary predator civilizations out there, and it's so quiet because most intelligent civilizations are smart enough to not go announcing their location. If you think about it, we've been pretty cavalier about broadcasting signals out into space and trying to contact alien life considering that we have no idea who is out there—and who or what might respond to our little messages.

Some of the greatest scientific minds of our times, including Carl Sagan and Stephen Hawking, were both vehemently opposed to trying to make contact with extraterrestrial intelligences because of this exact danger.

In the always poetic words of Carl Sagan:

"The newest children in a strange and uncertain cosmos should listen quietly for a long time, patiently learning about the universe and comparing notes, before shouting into an unknown jungle that we do not understand."

Possibility #3: There's just one super-asshole.

It could be that there is just one super-advanced predator species out there that keeps tabs on other technologically developing species and basically exterminates them when they cross a certain threshold. Terrifying. Moving on.

Possibility #4: Aliens have been here, but it was a long time ago.

Earth is 4.5 billion years old, but we've only had the ability to meaningfully look for signs of other intelligent life in the galaxy for the last century or so—a tiny blip in the history of both humanity and of the planet. Just because we don't see extraterrestrials hanging out around our cosmic neighborhood now doesn't mean that they've never been here—or that they won't return.

Possibility #5: We live in the boonies.

It could be that the galaxy is colonized, but that we just live in some far-flung outpost that they rarely visit. Basically the Earth could just be in a very unfashionable zip code. In all seriousness though, our solar system actually is way out on the edge of one of the Milky Way's spiral arms. Maybe we're just too far out to get visitors very often.

Possibility #6: Our technology is too primitive to detect them.

Just like you wouldn't be able to talk to someone on a cell phone with a messenger pigeon, it could be that the technology that is out there is just so advanced and so different that we're basically looking for the wrong evidence with the wrong tools. We assume that the signatures of intelligent communication would look the way that ours do, but we could be very wrong about that.

Possibility #7: We are too primitive to detect them.

In the wise words of renowned physicist Michio Kaku:

"If ants in an ant hill detect a 10-lane superhighway being built near them, would they understand how to communicate with the workers? Would they assume that the workers communicate only on ant frequencies? In fact, the ants are so primitive that they would not even understand what a 10-lane superhighway was."

It could be that we're ants in this scenario, and that extraterrestrial civilizations are right in front of us, potentially even sharing space with us, and our perspective from our current stage of evolutionary development is such that we don't even comprehend what we are seeing.

Possibility #8: The government is hiding the evidence.

We're going to talk about this one a lot more later, so we don't need to spend much time on it, but it is a possibility.

Possibility #9: The Earth is a zoo.

It could be that we are being observed by another species as a part of some sort of experiment. Or it could be that the solar system is being treated like some kind of a protected national park. Either way, there might be a "look, but don't communicate or interfere" policy with regards to our planet—sort of like the Prime Directive from *Star Trek*.

And if you feel like we're starting to get into Crazy Town with these possibilities, you should know that some of the most respected names in ufology and the disclosure movement—including Lue Elizondo—do not scoff at this idea and throw it around as a genuine possibility.

Possibility #10: We are fundamentally wrong about the nature of our reality.

In recent years, scientists have increasingly discussed the possibility that we could be living in a simulation. For real. So hypothetically, it could be that there just weren't any other advanced intelligences programmed into the simulation.

In fact, regardless of whether or not it helps us explain the Fermi paradox, the simulation hypothesis still might end up being true. The more we learn about quantum mechanics, the more evidence we find that could point to that startling conclusion.

A lot of those findings are too complex to get into now, but I'll give you this little thought experiment:

If you think about the evolution of computer and video games from the early 8-bit systems to the current emergence of virtual reality, it's clear that our ability to create more-and-more realistic virtual environments is increasing exponentially. Even with VR headsets now, it may not look real exactly, but your body still responds to it like it is. And with this exponential improvement of VR technologies it's not hard to imagine that at some point we're going to be able to create simulated experiences and environments that are indistinguishable from reality.

From there the logic is that if you could create this sort of simulation, you would. And hypothetically someone could create a simulated universe and in that simulated universe a species could advance to the point that it creates its own simulation within the simulation. And so on. Hypothetically there could be a near infinite number of simulated universes within the real universe, so if that's the case, then what are the chances that we're actually in base reality?

And that's actually one of the least convincing arguments for the simulation hypothesis. But for now, we're still on the alien question, and through the lens of the Fermi Paradox—and its many, many potential explanations—we can start to see how complex this question is, and how many questions we still have to answer.

Belief In Exterrestrial Life

As a kid born in the 80s, when I was growing up the idea that we were alone in the Universe was treated as the default and the most likely option. And I'm talking completely alone—as in, the conditions for even the most simplistic and rudimentary forms of life are so rare and mysterious that the Earth was likely the only place that it ever happened.

But as I've gotten older—and particularly in the past ten years—it feels like a shift has occurred where more and more people believe that the Universe is likely teeming with life, and it's only a matter of time until we find it. And we're even taking a long hard look at places within our own solar system.

I did some digging and it turns out that there have actually been some pretty dramatic swings throughout history when it comes to our belief about the existence of life on other planets. However, by and large, the dominant belief through most of recorded human history has been that the existence of life elsewhere in the Universe wasn't just a possibility, but a probability. Our more cynical outlook on the prospect in recent decades is more the exception than the rule.

These days our outlook on the possibility of finding extraterrestrial life is much sunnier. Two-thirds of Americans say that their best guess is that intelligent life exists on other planets.

What Extremophiles Tell Us About Extraterrestrial Life

To find evidence of the possibility of life existing elsewhere in the cosmos, we don't need to look any further than our own planet.

As we discussed with the Fermi paradox, until we find evidence of life somewhere else, it's hard to know how common it is for life to develop on a planet—especially considering that we don't actually know what causes life to develop in the first place.

We do know that our planet appears to be uniquely calibrated to support life, so we can probably assume that if there are certain conditions that are particularly favorable to the spontaneous creation of life they probably existed on Earth about 3.7 billion years ago. But how many other Earth-like planets might there actually be out there?

As we continue to identify exoplanets outside of our solar system, we're getting more and more data to help us answer that question. Although there is considerable debate about what exactly constitutes an Earth-like planet and how likely a planet that is like ours would be to develop life, evidence suggests there may be as many as one Earth-like planet for every five Sun-like stars in the Milky Way alone. That gives us a lot of chances for life to develop.

But what about places that are nothing like the Earth? Could life possibly arise in environments that aren't as friendly to life as we know it?

The more we learn about extremophiles the more we think that the answer could be yes. Extremophiles are organisms that live in conditions that we would normally consider to be hostile to life including extreme temperature, acidity, alkalinity, or chemical makeup. These organisms challenge everything that we thought we knew about what it takes to support life.

For example, there is a bacterium called *D. radiodurans* that not only thrives in the cores of nuclear reactors, but it can survive exposure to everything from toxic chemicals and corrosive acids to extreme heat above the boiling point of water, subzero temperatures, and the vacuum of space.

There are also three species of fungi that were discovered inside the abandoned Chernobyl Nuclear Power Plant. Scientists studied these fungi and found that they grow faster in the presence of radiation, and even effectively "eat" the radiation using the pigment melanin that captures the radiant energy, similar to the way it absorbs UV radiation in human skin to help avoid sunburns.

These extremophiles force us to rethink what it even means for something to be "alive" and to reconsider all of the vastly different ways that life might adapt itself to various environments. It even begs the question of whether or not the Earth really is uniquely suited to support life, or if we just view it that way because it is the only environment we know of that supports life as we know it.

There's even evidence that some extremophiles on Earth could survive in the conditions on Mars. So it's not difficult to imagine that there might be a wide variety of environments that could potentially give rise to at least simple life forms.

Wait...Did We Already Find Life On Mars?

And speaking of Mars, the idea that there could be life on Mars could be more than hypothetical. In fact, former NASA scientist, Gilbert V. Levin, claims that evidence of life on Mars was found back in the 1970s. When the Viking landers were sent to explore the Martian surface, Levin was the principal investigator on an experiment that he claims detected microbial respiration in the Martian soil.

He claims his team actually recorded four separate positive results, supported by five varied controls, coming from each of the twin Viking spacecraft which landed around 4,000 miles apart. For Levin, finding what was a clear byproduct of life was incontrovertible proof that there must be some sort of microbial life in the soil of Mars.

However, because actual organic matter was not found, NASA dismissed the positive results as being evidence of some sort of inorganic matter that was mimicking life, but was not life. No other explanation was offered for what that might be. In other words, "Don't worry about it."

Other Candidates For Life In Our Solar System

Still, it does make sense why NASA would hold off until finding actual life itself before making such an announcement. The profound implications for finding even microbial life on other planets is hard to overstate.

For one it would eliminate most of the answers to the Fermi paradox that are based on the idea that we are special. If our nearest neighbor, who we previously thought was potentially too hostile for life, proves to have microbial life, it would mean that we weren't that special and that the Universe is likely teeming with life.

So for now, we continue to look for signs of life in our solar system. And it turns out that we actually have some pretty strong candidates.

Europa

Europa is the smallest of Jupiter's Galilean moons—named for Galileo who first discovered them in the early 1600s—and is slightly smaller than our own moon. Ice-covered Europa might not seem like a great candidate for finding life at first glance, but due to an elliptical orbit around Jupiter that causes geothermal activity within the moon itself, there is believed to be an ocean of liquid water beneath the surface of Europa. This along with a thin, oxygen atmosphere gives scientists hope of one day finding signs of life in Europa's icy depths.

Titan

Although Titan is a moon of Saturn, it actually has more in common with the Earth than it does with other moons in our solar system—and, as far as we know, it's the only other place in the solar system where stable bodies of liquid can be found on the surface. Titan has many other features that would be familiar to us including a seasonal climate, weather, and familiar surface features like rivers, dunes, and deltas.

However, that's where the similarities end. Unlike Earth, Titan's atmosphere is made of nitrogen, not oxygen. And instead of oceans and rivers of water, Titan's liquid bodies are made of liquid methane and ethane. That combined with recorded temperatures of -290 degrees Fahrenheit, and we have to assume that if life does exist on Titan that it looks very different from life as we know it.

Enceladus

Another moon of Saturn, Enceladus, may also be harboring alien life. Like Europa, Enceladus has an icy crust covering subsurface oceans of liquid salt water. Tectonic activity inside the moon causes cracks and rifts in the surface ice, causing huge geysers that spray out of the moon, the trail of which helps to form one of Saturn's rings. And in those geysers scientists have been able to detect evidence of organic compounds that make up some of the building blocks of life here on Earth.

Is Interstellar Travel Even Possible?

So while we still don't have definitive answers about the existence of life on other planets, based on the existing evidence, it appears that our chances of finding life elsewhere in the cosmos is pretty good. Now whether any of that life could potentially be intelligent life is another question entirely.

But let's assume for a minute that there was intelligent life on a planet in another solar system. Would it even be possible for them to get here?

There are many who believe that, while there very well could be intelligent life on other planets, that the vast distances between solar systems would make it impossible for any species to visit even their nearest neighbors. And this isn't an insignificant problem to overcome.

Let's take our closest star, Alpha Centauri, and imagine that there is an Earth-like planet around it that we'd like to visit. It takes light from Alpha Centauri about 4.3 years to reach the Earth. Obviously, we'd need to travel much slower than the speed of light. So how might we get there and how long would it take?

Well to start with, rocket fuel—which we've traditionally used for space travel—is completely out. The fastest manned vehicle ever was the Apollo 10 rocket which reached a blistering speed of 25,000 mph. However, at that speed it would take 120,000 years to reach Alpha Centauri. Obviously, that's not ideal.

To get to Alpha Centauri in one human lifetime, we'd need to be able to travel at one-tenth the speed of light. But to travel that fast with rocket fuel, we'd need a fuel tank that is the size of the known Universe. That's definitely not going to work.

What we need is a much more efficient fuel than rocket fuel. So what else could we use to achieve interstellar travel?

Fusion

We could use fusion. Or put more directly, we could basically blow up a bunch of nuclear bombs behind us and ride that wave to another solar system.

Specifically, we'd need about 300,000 one megaton hydrogen bombs that we'd set off behind us, one at a time, at regular intervals for a month straight. That would be enough to get us to one-tenth the speed of light, and to Alpha Centauri in about 44 years.

However, it's not just enough to arrive at Alpha Centauri, we actually need to be able to stop—and stopping in space is not easy. In order to stop using this method, we'd need just as much energy to slow down as we would need to get up to our cruising speed, which means that we'd need to double the amount of time that it would take to get there.

So in this scenario it will actually take us about 90 years to get to Alpha Centauri, so we'd probably need to take three human generations on that trip to ensure they had enough people to man the ship for their entire journey. Suddenly this is getting pretty complicated.

So are there ways we could go faster? After all, although nuclear fusion is significantly more efficient as a source of energy than rocket fuel, it's still not very efficient. Fusion only turns about 1% of its rest mass into energy.

Antimatter Drive

If we wanted something that would turn close to 100% of its rest mass into energy, we could use a theoretical device called an antimatter drive.

Now, I'm not even going to attempt to explain what antimatter is. The top minds in science are still working on figuring that one out. Just know that antimatter is like the weird Wario version of matter—it is matter's exact opposite. So if you have one electron of matter, you could also have another particle that is identical to this electron, except with an opposite charge, called a positron. And if an electron and a positron collide they completely annihilate each other, releasing 100% of their energy.

So perhaps antimatter drives could one day be the answer to unlocking interstellar travel for humanity. But for now, it costs $100B to make even one milligram of antimatter, making it far too expensive to be a possibility until we find a way to create it at scale.

Light Sails

There's another hypothetical device we could potentially use to travel the cosmos called a light sail. A light sail would essentially allow us to harness light in much the same way that a sail harnesses wind—which would have the added benefit of meaning that we wouldn't have to weigh ourselves down carrying a bunch of fuel.

So how does it work?

Let's say that you have a kilometer wide sail that is coated in a reflective and heat resistant substance like sapphire. If you blasted it with a giant laser with the power equivalent of 100 nuclear plants, the sail would catch that light like wind and could conceivably get us up to one-tenth the speed of light. Hypothetically, we'd only be limited by the size of our sail and power of our laser.

However, the scale of such a project would be truly mind-boggling considering that the laser would be so big that we'd likely have to build it on the moon or we wouldn't be able to get it up into space. And it would have to be powered by massive reactors, which add plenty of complications of their own.

Black Hole Drive

Another hypothetical device we could use would be a black hole drive, which would involve creating a tiny black hole and harnessing its power. In this case, the black hole would be created with light and not mass. If you had a sufficient density of light concentrated into a small enough area—we're talking the equivalent of 600 billion kilograms, or two Empire State buildings down to the size of a single proton—it could bend the fabric of spacetime enough to create a singularity, and therefore a black hole. This black hole would radiate Hawking radiation, and the smaller the black hole, the more it would radiate, and hypothetically, we could use that energy to travel the Milky Way.

Wormhole

One of the top candidates for interstellar travel doesn't concern itself with next-level propulsion systems at all, because with this method, hypothetically anywhere in the Universe would effectively be right next door—because we'd be traveling through a wormhole.

I won't even attempt to get into the physics of wormholes right now, but all you really need to know is that wormholes are basically tunnels between two black holes that connect two different places in spacetime. Obviously this would be ideal because we wouldn't have to deal with figuring out a way to travel across mind-bending distances—but there are a few issues with this method.

The first is that technically, we don't even know for sure that wormholes exist. It *should* be possible, but until we actually manage to find one or make one, it's all just speculation. And the second is that, even if they do exist, we don't know if it would be possible for humans—or anything else, for that matter—to travel through one. And I don't know about you, but I don't think I'd volunteer to be the first one to try it out.

So, Is It Possible?

Going back to the question of whether or not interstellar travel is possible, the answer is a hard maybe. We can't do it now with our current technology, but we're at least close enough to be able to speculate about potential methods.

And it seems likely that we'll eventually be able to make it happen—though whether that will be in 100 years or 1000 years or 10,000 years is hard to say.

But certainly, if we think about an advanced civilization, especially a Type 2 or a Type 3, it seems almost a foregone conclusion that a species that had achieved that level of technological advancement would be able to crack interstellar travel.

So who's to say that they couldn't come here—or that they haven't already?

And What About Alien Abductions?

There's another pretty major piece of evidence that I, frankly, don't totally know what to do with—and that's reported cases of alien abductions.

It's one thing to talk about the likelihood of the existence of intelligent life in the abstract, or even to have more "nuts and bolts" conversations about the craft themselves. But as soon as you start talking about actual human encounters with a non-human intelligence, things get even more complicated.

Consistent Inconsistencies

One major complicating factor is that alien abduction is usually an individual experience. Even when multiple people claim to have been abducted at the same time, their experiences often differ in ways that range from minor to significant.

This also happens a lot with UFO sightings. For instance, take the Phoenix Lights incident, which was witnessed by hundreds of people over 3 hours and over 300 miles from the Nevada line down through Tucson. When you have that many people reporting this sighting independent of each and at the same time, it's hard to deny that they saw something.

But even among those witnesses there is a lot of variability in terms of what they saw. Some people saw 4 lights. Some saw five or six. Some people said the lights were white, while others said that they were orange or red. And while many people agree that the lights were part of one solid, v-shaped or triangular craft, others report seeing the lights moving around independently of each other.

This is pretty typical of reports related to the UFO phenomenon. There's an infuriating variability and lack of consistency not just between different sightings, but even within each of the individual sightings themselves. And these inconsistencies become even harder to parse when you're talking about incidents that involve just one or a handful of people, as is typically the case with abduction reports.

Could Alien Abductions Actually Be Sleep Paralysis?

Further complicating the issue is that many of the hallmarks of alien abduction reports sound remarkably similar to a relatively common phenomenon that nearly half of all people experience at some point in their lives—sleep paralysis.

When you enter REM sleep, a combination of neurotransmitters in your brain switches off motoneuron activity effectively paralyzing you, which keeps you from harming yourself or others as you dream. When this mechanism in the brain doesn't work correctly, people can end up sleepwalking, which can be dangerous. In extreme cases, sleepwalkers have been reported to do things like jump out of 5th floor windows, walk into traffic, or even commit murders.

In many ways, sleep paralysis is the opposite of sleepwalking. During normal sleep, the paralysis of REM sleep is basically switched off before you wake up. But when this doesn't happen correctly, a person can wake up in a state somewhere between sleeping and waking up and find themselves still unable to move.

When this happens the amygdala—the part of your brain that deals with identifying threats—freaks out and triggers a "fight or flight" response that causes feelings of extreme fear in the sleeper. And in this state of semi-wakefulness, people can still have dream-like hallucinations, which often take on terrifying forms because of the terror that the person is already experiencing.

And while these hallucinations can vary wildly, there are a lot of similarities between common descriptions of sleep paralysis and reports of alien abduction. People often report awakening to find a malevolent intruder standing over them, feelings of extreme fear, feelings of being pulled or dragged out of their bed, and even experiences that sound very similar to the grotesque and often weirdly sexual experiments and procedures that people report enduring during an alien abduction.

So many people believe that alien abductions are just misidentified cases of sleep paralysis. And there's other evidence that this could be the case beyond just the similarities of the experiences themselves. A significant number of people who claim to have been abducted by aliens say that they have had this experience multiple times throughout their life. And these experiences often seem to run in families where multiple family members over multiple generations claim to have experienced abduction.

Sleep paralysis more or less follows this same pattern, with around 40% of people experiencing sleep paralysis at least once in their lifetime and a much smaller percentage of people—about 6%—experiencing multiple episodes of sleep paralysis over long periods of time. And there is even evidence now that sleep paralysis likely has a genetic component suggesting that it could run in families, just like abduction experiences.

However, one glaring problem with this theory is that not all alien abductees claim to have been taken at night while they were sleeping. So while sleep paralysis could offer a solid explanation for many alien abduction cases, it can't explain all of them. So what else could be going on here?

Abduction Blueprint: Betty and Barney Hill

Let's look at the example of what is largely considered to be the first reported case of alien abduction—the Betty and Barney Hill Incident.

In September 1961, married couple Betty and Barney Hill were driving down a dark and winding country road in New Hampshire's White Mountains. They hadn't seen another car for miles, but the couple started to notice a bright light in the sky that seemed to be following them.

The next thing they remembered was arriving home in Portsmouth at dawn. They were dirty, their watches had stopped, Betty's dress was torn, and there was two hours of the drive that neither of them could remember.

The Hill's sought the help of a psychiatrist to help them understand this terrifying and unsettling incident. And under hypnosis, they claimed to have recovered memories of being abducted by gray beings with large black eyes who took them into a metallic, disc-shaped craft that Betty said was roughly the size of their house. On the ship they were subjected to an examination before their memories were erased and they were returned home.

In many ways, this story has become the blueprint for alien abduction. Despite the stunning variety of differences between abduction stories, there are certain elements of the Hill's story that pop up again and again—a bright light in the sky, a metallic and often disc-shaped craft, gray beings with large black eyes, failure of clocks and other machinery and electronics, missing chunks of time, and being the subject of examinations and experiments.

And also like with the Betty and Barney Hill incident, many abductees are only able to initially access memories of these incidents through hypnosis—which complicates things.

Could Alien Abductions Be Implanted False Memories?

There is a lot of controversy around the use of hypnosis to recover lost or repressed memories. Many psychologists even question whether it's even actually possible to banish memories to your unconscious mind and only become aware of them under hypnosis. Because memories are entirely subjective—and there's no way to know definitively what another person remembers or does not remember at any given time—the scientific evidence for this phenomenon is dubious, at best.

And even if the skeptics are wrong and it is possible to entirely forget a traumatic event and then recover it under hypnosis, there is the added layer of complication that comes from the fact that there is considerable evidence that it's possible to implant false memories into a person's mind through hypnosis.

For example, in 2011, a woman named Lisa Nasseff of St. Paul Minnesota sued her former therapist, Mark Schwartz, and Castlewood Treatment Center saying that he had falsely implanted memories during hypnosis that made her believe that she had been part of a satanic cult while receiving treatment for anorexia.

Nasseff claimed that she had experienced extreme anguish and was isolated from her family and friends because Schwartz had made her believe that her family and friends were members of a satanic cult who had subjected her to ritualistic abuse—including forcing her to eat babies.

It wasn't until she left the facility and began to connect with multiple other patients who also believed that they had recovered memories of being abused by a satanic cult that Lisa began to question these recovered memories. After all, what are the chances that multiple people in this one small facility had this exact same experience?

This questioning eventually led to the lawsuit by Nasseff, along with lawsuits by three other unnamed women—including one who had come home from the treatment center with memories of horrific sexual abuse by a neighbor that her family says logistically could not have occurred. Dr. Schwartz and Castlewood Treatment Center denied these claims. All 4 lawsuits were settled out of court two years later in 2013.

Many cases similar to this one have occurred over the past several decades, including the infamous Satanic Panic of the 80s and 90s. Many of these cases have been debunked, and even more have been called into serious question due to a lack of evidence to support the validity of the memories people have claimed to have recovered. So it seems likely that, at least in some of the cases where people recover lost memories, there is something deeper going on.

Could this be the case with Betty and Barney Hill and other people who claim to have experienced alien abduction? Could their memories have been intentionally fabricated, or even implanted into their minds, during hypnosis? It's hard to say, but it is certainly possible.

And that's what makes it so difficult to come to any clear conclusion on the alien abduction phenomenon and what it really represents. It's a highly subjective and messy business to try to come up with an answer that satisfies all the twists and turns of this bizarre phenomenon, and likely there is no simple answer.

Would Aliens Be Anthropomorphic?

However there is one aspect of the alien abduction phenomenon that is especially compelling and also deeply relevant to the alien question and it's worth taking a moment to explore. And that is the fact that in the vast majority of alleged alien abductions cases, the aliens that abductees report seeing are all similar in one very eerie and seemingly impossible way.

They look like us.

Many abductees claim to have encountered gray, big-eyed beings similar to those reported by Betty and Barney Hill, often referred to as "the Grays". Some abductees report seeing tall grays. Others see short grays that are only about three feet tall. Some see reptilian beings, while others see tall, blonde human-looking beings. But in the vast majority of cases, these various alien species, despite their differing appearances, are still distinctly and undeniably humanoid.

By-and-large, they have two eyes, two ears, a nose, and a mouth. They have two arms and two legs. They walk upright. Their entire body has the basic schema of a human body.

And for many people, that right there is evidence that the alien abduction phenomena must be the product of sleep paralysis, false memories, or even a hoax—basically anything except for actual extraterrestrials. Because, when you think about it, what are the actual chances that beings from another planet would look so much like us?

If evolution happens as the result of a mind-bending number of random mutations over time, and then those genes are selected based on their ability to help a particular species better thrive in a particular environment, how likely is it that beings that look so similar to humans would evolve on a completely different planet?

Just think of how different a human is from animals that have evolved to live in the extreme conditions of the Mariana Trench. And yet if you go back far enough, we have a common ancestor. How different might a species look that evolved on a planet with different gravity, chemical makeup, atmosphere, temperature, etc.? It feels logical to assume that if there are extraterrestrial beings out there, they would almost certainly look vastly different than anything that we've ever seen before.

And yet, surprisingly, many scientists are beginning to come around to the idea that aliens might actually look more like us than we might think.

Evolution May Be Less Random Than We Think

According to Charles Cockell, professor of astrobiology at the University of Edinburgh in Scotland, extraterrestrial life might look "eerily similar to the life we see on Earth." In Cockell's book, The Equations of Life: How Physics Shapes Evolution, he suggests that a "universal biology" may exist and that lifeforms on Earth could provide us with a basic blueprint for the kinds of life we are likely to find elsewhere in the universe.

His basic premise is that, although the genetic mutations that fuel evolution are random, the particular mutations that are selected for through natural selection are not. These mutations are still subject to the law of physics and also are restricted by the conditions and materials necessary for life to thrive. And according to Cockell, these factors narrow the scope of evolution to a degree that it's likely that life would evolve in similar ways on another planet.

We can see examples of how this might work right here on Earth. Animals that live in cold environments usually have thick fur. Animals that live in the ocean are typically hairless and have long, sleek bodies perfect for swimming in water. It's obvious just by looking at the different species on Earth that certain environments and conditions tend to produce certain body types and characteristics.

And despite these many differences, there are certain distinct similarities that the vast majority of life on Earth seems to share such as bilateral symmetry. Bilateral symmetry basically means that you can draw a line down the center of an organism and it is the same on both sides. Bilateral symmetry is great for things like being able to walk or swim in a straight line, as well as accurate depth perception.

When you try to imagine an animal that has any kind of significant asymmetry and how it might operate, it's easy to see how bilateral symmetry could very possibly be necessary to the evolution of complex and intelligent lifeforms. So evolution really might not be as random as we think.

Convergent Evolution

More evidence to support this idea comes in the form of a concept called convergent evolution. Convergent evolution describes organisms that evolve similar features entirely independently from one another.

For example, sharks and dolphins look very similar despite sharks being an egg-laying fish and dolphins being a mammal—and their last common ancestor swam the seas 290 million years ago. That's 60 million years before the earliest dinosaurs first appeared on Earth.

Even more surprisingly, flight appears to have evolved separately and at four different times in history: in insects, bats, birds and pterosaurs. The mechanism of how these animals fly is virtually identical.

Stranger still, both octopuses and humans separately evolved camera-like eyes with an iris, a lens. and a retina—a very exact and complex set of structures, all of which are essential parts of an imaging device. And this happened despite our last common ancestor existing 750 million years ago.

All of this is to say that, although it might be counterintuitive to think that extraterrestrials could have similar features and body structures to humans as is reported by most alleged abductees, the more we learn about how evolution actually works, the more likely it seems that these beings could look very much like us.

So Are You Telling Me It's Aliens?

So, I've thrown a ton of information at you, and yet it probably feels like we're no closer to an answer when it comes to the question of whether or not UFOs could be extraterrestrial technology than we were when we started.

And that's OK. If we're going to continue down this rabbit hole, we need to get cozy with ambiguity and uncertainty, because there are no straight forward answers here.

We can say that it's probable that alien lifeforms exist, and that a certain percentage of that life—though we don't know how large or vanishingly small that percentage might be—has likely evolved to a level of intelligence that would make space travel possible.

We can say that interstellar space travel is theoretically possible, and that it seems likely that a more advanced civilization than ourselves could undertake such an enterprise.

And while we can't say with certainty that humans have yet encountered any of these beings, the reports that we do have of such alleged encounters are more or less in line with what scientists would expect these beings to look like.

But is it aliens? That's hard to say.

There's this idea in math called a transitive property that basically says that if a=b and b=c then a=c. So, for example, cats are mammals. Mammals are warmblooded. Therefore, cats are warmblooded. And this feels like a situation where you could say that if UFOs are real and if UFOs are alien technology then aliens must be real.

But that logical leap only works *if UFOs are alien technology*— and we're not at a place where we can assume that. In fact, as we'll discuss in the next two chapters, although we tend to equate UFOs with extraterrestrials, there are several other possibilities that could potentially explain the UFO phenomenon, as well.

Something Strange Is Going On

And, if you listen closely to what former AATIP Director, Lue Elizondo— and others in the disclosure movement with access to privileged information—are actually saying, you may begin to wonder if what we're dealing with really is of extraterrestrial origin.

Here was Lue Elizondo's response to a question on the *Fade To Black* podcast when host Jimmy Church asked a question about the origins of the infamous Tic Tac video from the Nimitz incident.

"Jimmy, I'm sorry for saying this. I don't want to correct you publicly, but you're asking the wrong question. The question should be, "Is the Tic Tac man-man?" or "Is the Tic Tac U.S. or foreign technology. The answer is no. But when you say, "Is it from this Earth?"—we don't know. That's why, you know, we've always stated before, from our position, we don't know if these things are from outer space, inner space, or frankly, the space in between.

As you begin to understand the world of quantum physics, there's a whole lot of reality around us and we don't interact with it. You know, we're used to the Newtonian understanding of the world where there's gravity and the apple falls from the tree. Then along comes Einstein a few years later who talks about a thing called relativity, where space-time is actually flexible, and only in the last 20-30 years have we started to explore the world of quantum physics—the quantum world, quantum states.

And we're realizing that all these realities, as opposite as they may seem to us, are actually very much a part of the natural environment. My point being, there may be realities all around us that we simply don't perceive. You know, we're very limited as species. We live in a three dimensional space with an XYZ axis and time which is the fourth dimension which is expressed in a linear motion for us, but that's not to say that there aren't other things all around us.

The UFO Rabbit Hole

We sense the Universe through the five primary senses of touch, sight, taste, smell, hearing, and yet we know that we only perceive 1% of 1% of the Universe that's around us. Most of the Universe remains completely invisible to us. And I don't mean the Universe "out there", I mean the Universe right here. Let me put that into context. We know 5% about our deepest oceans. We really don't have a good idea of what lives on our own planet, let alone anywhere else.

When you look at the human being, we are exactly in the middle of the scale of the Universe. If you were to take us, being the smallest thing, and the grandiosity of the Universe—which is 13.5 billion light years in this direction and 13.5 billion light years in the other direction—we seem pretty small. By that scale there is equally the same amount of space inside of the human body. There is this entire Universe inside of us, when you start to talk about just not the cells of the body, but the atoms—you talk about the subatomic particles and quarks—this is real, even though we can't interact with it. There's an entire Universe inside every one of us.

Now, I don't want to go too far off the beaten path here. My point being—there is so much we don't understand about the Universe, let alone our own selves. So I think it is foolhardy for us to even begin to say that the buck stops with us."

Crazy right? And then there was a puzzling and intriguing answer that he gave on *That UFO Podcast* when he was asked about a previous statement he had made where he said basically, "What if it's not *mankind* but *mankinds?*" Check it out:

"Boy, you're getting some really, really good and personal questions. Let me see if I can answer that in a meaningful way without leading the witness.

We live in a three dimensional world. We live in a three dimensional world where time is a function of the fourth dimension, if you will. And we experience time as being linear. It's a linear function. But we now know in the quantum world that—

First of all, we know that space and time—because of the work of Einstein and others—we know that space and time are joined together. And we know that space-time is flexible. It's called relativity. And we see it around massive objects all the time. Not just the Earth, but the Sun, black holes. So the linear Universe that we experience, to some degree really isn't.

When you look at the quantum state of things—even an electron, for example— you learn in high school that an electron orbits an atom. We now know that is a very simplistic way of looking at things. The electron doesn't orbit around the atom. In fact, it's called an electron cloud because the electron is both in all places and none of the places at the

63

same time. It's kind of weird, but—

My point being, that if you were to look at time—think of a cigarette burning. When you look at the notion of the future, most people would define the future as 'those events that have not yet happened'. And the past is defined by events that have already happened. And when you look at that construct then, the definition of the present must be a moment in space-time—probably measured in Plank time, a very infinitesimally small moment of space-time—where the future is transitioning into the past.

It's not a point in time—it's a process, it's an event that's occurring. So a way to put that in lay perspective is think of a cigarette or a cigar. Those parts of the cigarette that have already burnt to ash are the past. That part of the cigarette or cigar that has not yet burned is the future. And the cherry—that moment of ignition, that spark where the future is being consumed and is becoming the past—that is the present.

And we as human beings, we live all of our lives in the present. All of our hopes, our fears, memories, love, hate, good, bad—all that is an expression of an experience that occurs at an infinitesimally small moment of space-time. Like I said, probably measured in Plank time. That is how we experience the present. |

But what if there were things that had the ability to experience things where the present was a much bigger cherry, if you will—a much bigger transition? More elements of the future and the past are experienced as being in the present, and can do that also physically? So it's not just an idea, but what if there were species out there that experience the Universe with an extra level of dimension?

And so, you and I are having this conversation right now with your audience, and we're having this conversation right here, right now. But if I were to have the ability to have this conversation right here, but five minutes ago, or five minutes from now, we would never meet. We'd be like two ships passing in the night. Is it possible that some of these things, these UAP, have this ability and we experience them when they are right here, right now, and every other time we don't because we're simply not intersecting with that extradimensional space of time?

When you look at that cigarette or cigar burning, you'll notice that it never burns evenly. When you look at it up close, you can kind of remove the glare and the flare from the burn and we realize that there are portions of the future—portions of the cigar that haven't burned yet—behind portions of the cigar that already have burned. It's not an even burn. There's an overlap.

And quantum theory is beginning to show some of the models for that. So, I know I'm going in a very long-winded, round-about way to answer your question about what I meant by "mankinds", but I guess my answer to that is that it's limitless. Every time we put a limitation on Mother Nature she defies our box. "

So that's where we will pick up next time, with the idea that the extraterrestrial hypothesis for the origin of UFOs, is one of the *less* bizarre and exotic explanations. And what we are dealing with here might lead us to answers that shatter our paradigms in ways that could make us wish for something as simple and clear-cut as an alien invasion.

4 ARE UFOS HUMANS FROM THE FUTURE?

Over the last two chapters we've discussed two potential explanations for the UFO phenomenon—that the highly advanced technological craft that are being recorded in our skies and in our oceans could be secret human technology, or that they could, in fact, be extraterrestrial.

But despite the fact that we tend to associate UFOs with either of those two possibilities, we're just barely scraping the surface in our understanding of what these craft could actually be. And I've got to warn you—from here on out, things get pretty strange.

In the last chapter we talked about one of the most puzzling aspects of alleged alien encounters—which is that they look a lot like us. In other words, they're anthropomorphic. Which is a mashup of two Greek words— *anthrop* meaning "human" and *morphic* meaning "shaped"—so literally human-shaped.

And as we discussed, although the idea that extraterrestrial beings would be anthropomorphic is counterintuitive, the more we learn about how evolution works, the more likely it seems that we could encounter intelligent beings from other planets who look a lot like we do.

The Future Human Hypothesis

But what if there was another explanation? What if they aren't just human-shaped, but actually human? What if what people are seeing aren't extraterrestrial beings that happened to evolve along a similar evolutionary path—what if they are a *part* of our evolutionary path, but they're just further down the road than we are?

This is what is referred to as the Future Human Hypothesis. The basic premise of this hypothesis is that the origin of the UFO phenomenon isn't from another world, but from this world far in the future.

Michael Masters, a professor of biological anthropology at Montana Technological University is a proponent of the future human hypothesis. In his book <u>Identified Flying Objects: A Multidisciplinary Scientific Approach to the UFO Phenomenon</u>, Masters makes the argument that, although we tend to associate UFOs with extraterrestrials, that it actually makes a lot more sense that they have a terrestrial—and more specifically, human—origin, saying:

"We know we're here. We know humans exist. We know that we've had a long evolutionary history on this planet. And we know our technology is going to be more advanced in the future. I think the simplest explanation, innately, is that it is us. I'm just trying to offer what is likely the most parsimonious explanation."

And he does have a point. As we discussed in the last chapter, until we get incontrovertible proof of extraterrestrial life, we just simply don't have enough data to know how common it is for life to develop and how often that life evolves to a level of technological advancement that would make interstellar travel possible. And even then, interstellar travel is something that is only hypothetically possible using hypothetical technology. There are a lot of unknowns there.

And as Masters points out, the future human hypothesis doesn't need to overcome any of those challenges. We don't need to unlock the mysteries of the universe to find an explanation for this phenomenon. All the answers we need are right in front of us.

And yet, as we'll see, the future human hypothesis offers us anything but easy answers. So let's take a little time to explore this idea, starting with the question of how humans might evolve to look in the future.

How Will Humans Look In The Future?

It's impossible to play the tape forward on evolution and know for certain what humans will look like in the future, but we only need to look at our own evolutionary path up until this point to see how drastically a species can change over time. And we know enough to make some basic assumptions that allow us to at least approach this as an interesting thought experiment.

So, based on what we know, what might future humans look like?

Well, if current trends continue, they will probably look much more similar to each other than present-day humans do. Humanity, as we know it, is very diverse. Over the last 200,000 years or so, as humans spread out across the globe, we developed different traits based on adaptations to our environment, interbreeding with other ancient human species, and random mutation. As a result, there is a stunning amount of variety in our skin color, eye color, hair color and texture, height, etc.

However, as we become more technologically advanced our world is getting smaller. We're no longer isolated from each other in little pockets of humanity, and we're increasingly mixing together in one giant genetic melting pot.

The natural result of this is that our genes are becoming more homogenized. Dominant traits like brown eyes become much more common than recessive traits like blue eyes. And the traits that aren't tied to one specific gene like skin color and eye shape blend together causing people to look more and more alike.

But aside from looking more alike, how else might humans evolve physically in the future?

In 2013, artist and researcher, Nickolay Lamm, worked with computational geneticist, Dr. Alan Kwan, to illustrate what humans might look like 100,000 years from now. Based on their research they theorized that human foreheads would continue to get larger and cranial capacities would continue to accommodate larger brains—a trend that started in the 14th century.

They also proposed that as humans begin to colonize space, that we will probably have much bigger eyes to help us see better in low-light environments. According to Kwan, these eyes would likely be "unnervingly large" by today's standards. And we might even develop eye shine to further enhance our low-light vision.

Not everyone agrees with these predictions—and honestly, there's no real reason for them to do so. It's all just wild speculation.

What it comes down to is this: close-encounter accounts typically describe UFO tenants as bipedal, hairless, human-like beings with large brains, large eyes, small noses and small mouths. And that's interesting—super interesting. But just because the beings that people report seeing in and around UFOs look like us, doesn't necessarily mean that they are us.

But they could be.

What Might Future Humans Be Doing Here?

So let's have a little fun with this and assume that UFOs are future-human in origin. The first and most obvious question is what are they doing here?

There are three basic buckets of possibilities:

1. They're just curious.
2. They're trying to save us.
3. They're trying to save themselves.

They're Just Curious

Let's start with the idea that they are just curious. The thinking with this theory is that if a species evolved technologically to the point where time travel was possible, then the natural inclination would be to go back in time and check out what things were like in the past.

So it could be that they are basically tourists. Maybe in the distant future instead of taking a Caribbean cruise, you can vacation by going back in time to see dinosaurs, then fly through Egypt to watch the pyramids being built, and then pop up north and a few thousand years to catch the fall of Rome before heading back home. That does sound pretty awesome. And it makes sense that if we could do such a thing, that we probably would. So that is one possibility.

Another possibility is that they could be more formally studying us. Time travel would be an absolute game-changer for historians and archeologists of the future. Instead of painstakingly collecting artifacts and data and piecing them together, scholars and scientists could go back and see what was going on during a particular time period for themselves.

They're Trying To Save Us

It also could be that they aren't here just to observe us, but that they are here on a very specific mission. And some people think that that mission could be to save us by preventing some massive calamity. Maybe they are here to warn us of an impending disaster. Or maybe their goal is simply to alter the timeline just enough to avert some unknown outcome.

One major reason that some people gravitate toward this particular explanation is that there is an undeniable connection between UFOs and nuclear technology. Where there are nuclear weapons and nuclear facilities, you tend to find lots of reports of UFO sightings.

Over the last decade or so, more and more current and former members of the U.S. military have come forward to talk about UFO encounters in which these craft didn't just seem to be interested in our nuclear weapons, but interacted with them directly.

One such alleged incident occurred on March 24th, 1967 at Malmstrom Air Force Base in Montana. U.S. Air Force Captain Robert Salas was the on-duty commander of a secret underground launch control facility that night and he claims that he saw a giant UFO disable ten nuclear weapons.

The saucer-shaped craft was seen floating above the base by multiple guards stationed on the ground. And Captain Salas watched in astonishment as all ten nuclear weapons were powered down and rendered useless— something that should have been impossible as all ten of the weapons were running on ten independent systems.

Salas reported that it took over a day to get them all back online. No damage to the weapons systems was found, nor has there ever been an explanation for what happened that day. And this is just one of literally over 100 reports from members of our military regarding this sort of phenomenon at nuclear sites.

Many people look at this correlation and wonder if it could be that future humans are trying to warn us off from the devastating potential consequences of using our nuclear weapons. And it's not like nuclear war is the only potential disaster that we're currently flirting with as a species. It could be that they are doing their best to save us from ourselves.

They're Trying To Save Themselves

And finally, it could be that they aren't trying to save us, but trying to save themselves. We might have something that they need to survive. But what could be going on in the future that would cause humanity to look back into its past for help?

Going back to Michael Masters and his book on the future human hypothesis, <u>Identified Flying Objects</u>, he proposed that they could be coming back to harvest our DNA. The idea is that if the homogenization of human DNA continues into the future, that could make us particularly vulnerable to extinction.

For a species to stay strong and viable, there needs to be a lot of variation among individuals. If there isn't enough variation, the species will become vulnerable to diseases caused by everything from recessive traits to viruses—and they could be wiped out.

And if you needed to introduce more variation into your DNA, your ancient ancestors would be a great place to start, if you had the technology to do so.

Once again, this is all purely speculative. But it's possible.

Ross Coulthardt and The Future Human Hypothesis

I'll be honest and say that I struggle with the Future Human Hypothesis. For reasons that we'll talk about in just a minute, I just ultimately don't find it to be very satisfying. However, there is one odd bit of potential evidence in support of this idea that I personally find too compelling to ignore.

There is this award-winning, investigative journalist from Australia named Ross Coulthardt who recently wrote a book on the UFO phenomenon called <u>In Plain Sight: An Investigation Into UFOs & Impossible Science</u>. This was a big departure and a big risk for someone who had built his career and reputation on being a "serious" journalist.

And yet, Coulthardt stuck to his guns, believing that there was more than enough evidence to not only confirm that the UFO phenomenon is real, but that there has been a massive international coverup to hide this fact from the public.

And recently, he's come out in interviews and said that, according to some of his high-level contacts inside the U.S. government, UFOs are, in fact, humans from the future. This is the answer he gave in an interview with Curt

Jaimungal on *Theories of Everything* when he was asked to comment on Lue Elizondo's now infamous "somber" comment.

Ross: *"I wish I could tell you what I'm being told right now, but I don't think it's responsible for me to talk about it until I've been able to verify it more—because I don't want to panic people or be irresponsible. And I've been told in another area, certain things about the phenomenon that are quite disturbing. I mean, there are a lot of people who are privately claiming to me things about the implications of the phenomenon, that go beyond—far beyond—*

I mean, I wish it was as simple as extraterrestrials getting in their little space ships and coming from Zeta Reticuli and flying to this planet. That's the easy explanation.

The explanation that I've been exploring in recent months is more complex and I've already spoken about this to some extent. So I will say that it involves the notion of future human time travel. And look—it's only hypothetical. I'm not saying it's real. But if what I'm being told about that is true—yeah. I would be somber, too."

Curt: *"Why is that somber? Why is the fact—or the potential—that it might be humans from the future terrifying?"*

Ross: *"Because of what it—well, I don't think I'd be giving too much away if I said that—you think about—*

Why, since 1947, has there been a phenomenon taking an interest in the human race, particularly in nuclear weapons? Why is something, or someone, apparently trying to send us a message about nuclear weapons? Why is it that nuclear weapons are being shut down by what Slide 9[1] referred to as 'remote sensor disassembly'? What's it trying to say? What's coming?

Imagine if—and I'm only speaking hypothetically here—imagine that if future humans knew that if we continue on the path we're on that there's going to be a nuclear war or a conflagration of some type. Wouldn't you want to head that off? Wouldn't you want to protect your kin?

And imagine if you were somebody in the U.S. defense and intelligence establishment that was aware of that quite insane sounding idea—that this might be a time thing, a future war thing. And imagine if you're worried that anything you do might

---

[1] Slide 9 refers to a controversial slide from an alleged leaked presentation from AATIP which appears to be from a briefing about their findings regarding UAPs and outlines abilities identified by the study.

jeopardize that time stream. That's where I'm working at the moment. And I don't know if I'll ever get an answer."

Wild, right? Although I'm not the biggest fan of the Future Human Hypothesis, Ross Coulthardt is someone who is well-respected, not just in the UFO community, but in the world of mainstream journalism. He has a reputation for being a hard-nosed and fearless investigator who has taken on and taken down everyone from terrorists to organized crime syndicates. So when Ross says something like this, it's hard not to listen and wonder if he might be onto something.

Granted—and Coulhardt himself is aware of this and is the first to admit it—his information is only as good as his sources. And right now we don't know who those sources are, or what their motivations might be for telling him these things. The reality is that intelligence agencies lie all the time, and it could be that he's being intentionally fed misinformation.

But Coulthardt's sterling reputation leads many to believe that this is very likely what he's being told by his contacts in the government. Whether or not it's true is still up for debate.

Problems With The Future Human Hypothesis

Because here's the thing—the Future Hypothesis has some real issues. There's not enough evidence to rule it out, but like I said—there's something about it that is just unsatisfying.

First of all, let's say that Coulhardt's sources are right and they are here to save us from some impending disaster. It begs the question, why wouldn't they just tell us that? Why would they instead decide to dick around for 80 or so years, popping up randomly but without any clear intention or message?

And also, even if there was a huge calamity of some kind, I don't know that I totally buy that whatever remains of humans in the future would choose to go back and fix it. Think about it: if you had the ability to go back in time and prevent some horrible atrocity from happening—like if you could go back and prevent the Holocaust from happening—the knee jerk reaction is that yes, of course, you'd do it. In a heartbeat.

But how far back would you go? And what would you change to be certain that it wouldn't happen? And—and here is where it gets messy—what impact would making those changes in the past have on the people who are alive in the present?

What if no one lost a loved one in the horrors of the concentration camps? What if all the high school sweethearts didn't have to go away to fight a war never to return? It sounds fantastic—like a miracle.

But then you consider all of the people who were born because those people never returned. If you go back and save them, how many people are you killing in the present? And is causing them not to exist the same thing as killing them? And what if you found out that saving millions of lives would mean causing millions more to never exist?

And how would you even make that decision? Do the people who were around first get priority when it comes to choosing who to save? Do you make the choice that saves the most lives or do you make the choice that causes the least amount of suffering? And how would you even know any of those things ahead of time?

There are no easy answers when it comes to anything that involves time travel, particularly when you start talking about altering timelines.

And that's really where the primary issues with the Future Human Hypothesis lies—with the idea of time travel itself. It causes all kinds of problems that we aren't sure how to solve. And much like anyone who claims to understand the plot of *Tenet*, I'm suspicious of anyone who claims to have any definitive answers on this issue. Because, to start, we don't even know for sure that time travel is even possible.

Is Time Travel Possible?

Actually, that's not true. There is one kind of time travel that we know is possible, but unfortunately, it doesn't really help us with this scenario. And that's because, the only kind of time travel that we know how to do, at least theoretically, is going forward into the future—but with this kind of time travel, it's a one-way ticket.

Traveling Forward In Time

We know that we can go forward in time through a property of Einstein's Theory of Special Relativity called time dilation. So, as a quick refresher, we know from Special Relativity that time is not constant, and that the faster you move—or the stronger the gravitational field you're exposed to—the slower time moves for you.

In a technical sense, all of our astronauts who've spent any time in space are also time travelers because as they move at tremendous speeds in space, time is actually moving fractionally slower for them than it is on Earth. However, the difference is only about 38 microseconds per day, so we don't notice it.

But if we were to be able to put a human into a craft that could travel at near the speed of light, it would be possible to send that person far into the future, while only a few hours or days pass in their time. But this only works when you are trying to go forward in time.

So obviously that's a problem. Because while it could give us clues as to how we might be able to visit future humans, it doesn't explain how they could come see us. And if the trip is only one-way all of the primary theories for why future humans might come here start to fall apart.

Traveling Backward In Time

So let's go back to this idea of traveling backward in time. Sure, we can't do it now, but if we're talking about human technology thousands or even hundreds of thousands of years in the future, it seems like we could probably figure it out eventually, right? So why are so many scientists adamant that traveling backward in time isn't possible?

The main reason is the Second Law of Thermodynamics. Almost all of the laws of physics are symmetrical with regard to time, which means that they would work the same way whether time were going backward or forward.

For instance, imagine that you're playing pool. You hit the cue ball with your pool cue, and that cue ball hits the 8 ball into the corner pocket. You win! Hooray. And if you wanted to understand exactly how much force was transferred from your hand to the pool ball to the eight ball, you could use physics to figure that out. It's just a math equation.

But if you could run time backward and watch as the eight ball pops back out of the corner pocket to roll back and hit the cue ball which rolls back to hit your pool cue, you'd basically have the exact same math equation, but just in reverse. If all of the laws of physics were symmetrical in this way, technically there would be no real reason for time to move forward instead of backward.

However, the Second Law of Thermodynamics is the exception. This Law states that over time, everything moves from an ordered state to a disordered state. This is called entropy.

Think about what would happen if you put a drop of ink into a glass of water. Very quickly the drop starts to lose its shape. It spreads out through the water, and within a few minutes it will have lost its shape and structure entirely. You won't see a drop of ink in water anymore. The water will be dyed one, consistent color as the ink distributes itself evenly within the glass.

No matter how many times you try it, the ink in the glass will do this in a couple of minutes, totally on its own, even if you don't stir it. And yet, no matter how long you leave that glass there—even if you could somehow leave it there undisturbed for 1000 years, the ink molecules would never come back together. Once the ink has distributed itself through the water, you can't get the ink back out again.

The Second Law is why you can mix the ingredients for a cake, but you can't unmix them. And this doesn't just apply to cakes, but to everything in the Universe. Once something is mixed up, it can't be unmixed. This implies a directionality to time that many physicists think can't be overcome.

Can You Change The Past?

Granted, not everyone agrees, and there are some physicists who think that backward time travel could be possible, but most of them agree that even if you could travel backward in time that you probably wouldn't be able to change anything.

The example that's often used is the Grandfather Paradox. Basically, if you went back and killed your grandfather before your father was born, then you would never be born. But if you were never born, then who killed your grandfather?

Many of these physicists suspect that if you went back in time and tried to change things that you would be prevented from doing so in some way. Like let's say you went back in time to try to prevent the COVID pandemic by stopping Patient Zero from getting infected. The theory is that even if you were able to stop that from happening, that the pandemic would still happen in some other way. Maybe you would become Patient Zero.

It's all highly speculative, because—while it's fun to think about—time travel doesn't just break our brains, but it breaks everything we know about the laws of physics and how they work.

So that's where we'll leave the Future Human hypothesis for now, and in the next chapter we'll explore yet another potential explanation for the UFO phenomenon—yes, there are more. And as we go further down this rabbit hole, the ideas that we will encounter grow curiouser and curiouser.

5 THE INTERDIMENSIONAL & ULTRATERRESTRIAL HYPOTHESES

I hope everyone is well-hydrated and has done their stretches, because this chapter is where we start a whole new leg of our journey. This, my friends, is where we start to truly go through the looking glass, exploring the shadowy region of human experience that lives beyond the boundaries of our understanding.

It's the place where the deepest mysteries of science collide with the world of the hypothetical, the mythical—and, for some, even the divine. And if I'm being honest, I'm actually a little bit nervous.

Because—while we've flirted with the fantastical throughout these last few chapters—we've done so while holding firmly to the guide rope of proven, or at the very least "respectable", science. But if we're going to go any further and truly begin to grapple with the mystery at the heart of the UFO phenomenon, we have to be willing to venture outside of the boundaries of the established narrative.

In other words, things are about to get pretty strange.

And, to be honest, there's something about doing this that feels like a risk. After all, there are things that we simply don't say aloud in polite society. There are topics that we just don't touch for fear of what will rub off on us. Getting to this point in our conversation and speaking about these things publicly feels like crossing the Rubicon. There's no going back from here.

I've really struggled with this intro over the past few days because no matter how many times I restart it or come at it from a different angle, it just keeps coming out sounding like an apology. Which I hate. Because I'm not sorry.

Listen—something is happening here. Something big. And there are a million things that we don't know—and maybe can't know—about what it is or what it means. But what is clear is that even the most mundane explanations for whatever this phenomenon might be would change everything about life as we know it.

When I started this line of questioning earlier this year, the thing that was most startling to me was the realization of just how certain I was of things that weren't certain at all, and how immediately dismissive I was of huge swaths of knowledge and evidence simply because it didn't fit into this respectable little box of ideas that an intelligent, college educated, white collar professional is allowed to entertain.

So I made the commitment to myself—quietly, because I was still worried about what people would think—to go back to the drawing board and to learn everything that I could about this phenomenon. I would keep an open mind and do my best not to judge the data or let my own biases dictate my path, but rather just let it unfold and follow it wherever it may lead.

I'm still on that journey, and I still don't have the answers that I'm looking for, but it has been a truly life-altering experience to recognize how much I've let other people's opinions shape, not just what I believed, but what I even allowed myself to entertain. I've always considered myself to be a clear, rational, independent thinker—and it's been humbling to recognize the extent to which that has not actually been the case.

And to be clear, I'm not asking that you accept or believe any of what follows. All I ask is that you try not to reject it outright. To get the full picture of the UFO phenomenon, we need to look at every piece of evidence and every theory, even the ones that seem too outlandish or don't seem to fit. So when we come across a piece of information that seems ridiculous, or that makes you uncomfortable, or that you just don't know what to do with, just put it on the shelf

You can come back to it from time to time. Pick it up, reconsider it in the light of new evidence, and weigh it against different ideas that you've picked up along the way. And eventually you'll either find a use for it or you'll get rid of it. Or maybe you'll keep it on the shelf as a fun little oddity that you pull out at parties. But wherever it ends up, you'll be putting it there after thoughtful consideration.

You may very well still find a lot of what we talk about in this chapter to be preposterous when all is said and done, but in keeping an open mind on this journey, you'll be giving yourself more complete access to the full scope of thinking around this phenomenon, allowing yourself to form what you believe to be the best explanation on your own.

So with that in mind, let's get moving. We have a lot of ground to cover today as we consider some of the more exotic explanations for the UFO phenomenon.

Valleé, Hynek, & A New Way Of Approaching The UFO Phenomenon

And a fantastic place to start this conversation is with the work of Jacques Valleé.

Valleé is an internet pioneer, computer scientist, venture capitalist, author, astronomer, and one of the most venerated of Ufology's founding fathers. His work has, in many ways, defined modern ufology. And it's almost impossible to get a true handle on the more unusual hypotheses for the origin of the UFO phenomenon without understanding his work.

Valleé's interest in the phenomenon began in 1955 when he saw a UFO over his home just outside of Paris, France. Six years later in 1961, while working as an astronomer for the French Space Committee, he claimed to have witnessed the destruction of tapes of an unknown object orbiting the Earth.

The peculiar object was in a retrograde orbit—something that is rare among man-made, artificial satellites because of the difficulty of launching into orbit against the rotation of the Earth. Valleé and his team assumed what they had found was a natural satellite that had been captured by the Earth's orbit, but before they could investigate it further, an unnamed supervisor came into the room and erased the tape.

This sparked a lifelong interest for Valleé in the phenomenon. But despite his personal experiences and his lifelong dedication to searching for answers, Valleé is so well-respected because he has never been a fanatic, and has always been a scientist first.

Valleé received his PhD in industrial engineering and computer science from Northwestern University in 1967. It was there that he met his mentor, J. Allen Hynek, who was chair of the astronomy department, and the two began to conduct non-institutional research into the UFO phenomenon.

Hynek is an interesting character himself, and understanding his background can further illuminate Vallee and his work.

Between 1947 and 1969, Hynek served as a scientific advisor to the Air Force in their study of UFOs under the codenames Project Sign, Project Grudge, and Project Bluebook. Although the name changed over the years, the mission of this project remained the same—to investigate, and most importantly, to debunk reported UFO sightings.

In this role, Hynek was the Air Force's lead debunker, a job that the hard-nosed, highly skeptical scientist relished. He regarded the UFO phenomenon as "utterly ridiculous" and as a "fad" that would soon pass. Much of the language that is used to this day to discredit and stigmatize the phenomenon and those who report having experienced it is a direct result of Hynek's intentional and systematic approach to the subject—including being the first to attribute UFO sightings to "swamp gas".

However, after two decades of working with the Air Force, Hynek changed his tune and began to study the UFO phenomenon in earnest. In 1973 he founded the Center for UFO Studies which was dedicated to advocating for the scientific research of the UFO phenomenon—a cause he championed, both at home and abroad, until his death in 1986.

Hynek attributed this change to two things:

The first was what he called "the completely negative and unyielding attitude" of the Air Force in their approach to studying the phenomenon. In his words, "They wouldn't give UFOs the chance of existing, even if they were flying up and down the street in broad daylight."

And the other was the caliber of some of the witnesses that he interviewed, including members of the military and police officers—people that he considered to be trained observers. He also had first hand knowledge of how such reports were received by higher ups in those organizations, and that those coming forward were risking not just ridicule, but their careers.

I tell you all of this, partly because, as with any complex topic, understanding the history of ufology is important to understanding the topic as a whole. But also, I think it's important to recognize ufology as a legitimate field of science that has been pursued with both great passion and with great scientific rigor by some of the best and brightest minds within some of our most well-respected academic institutions—even if it has rarely been embraced by the institutions themselves.

And like Valleé, Hynek wasn't just another breathless believer chasing down any evidence that might confirm his preexisting beliefs and biases. He was a scientist, a skeptic, and someone who spent decades debunking and explaining away the phenomenon—something that, at the time, he truly believed was in the best interest of the public. And despite his change of opinion on the topic, he brought the same level of scientific discipline to his research throughout his career.

So circling back to Jacques Valleé—

Although he had originally begun his research into the phenomenon with the goal of validating the extraterrestrial hypothesis, by 1969—right around the time when he started working with Hynek—Valleé began to express doubts about its viability stating that the extraterrestrial hypothesis was too narrow and ignored too much data.

Valleé's 5 Arguments Against The Extraterrestrial Hypothesis

By 1991, Valleé crystalized this argument in his paper published in the *Journal of Scientific Exploration*, titled *Five Arguments Against The Extraterrestrial Origin Of Unidentified Flying Objects*, which are as follows:

The Reports Are Too Numerous

The first argument is that unexplained close encounters are far too numerous to be simply visitation or study by extraterrestrial visitors. Valleé was one of the first researchers to attempt to compile reports of unexplained aerial phenomena from around the world, and was shocked to discover over 900 reports spanning the 100 years between 1869 and 1969—which he published in his groundbreaking book, <u>Passport To Mogonia</u>, that same year.

But by the time he published his "Five Arguments" paper in 1991, anywhere from 3,000 to 10,000 of these reports had been collected, depending on which selection criteria one used to decide what constituted a legitimate report. Valleé offered a relatively conservative number of 5,000 reports. And arguing that, due to factors ranging from the stigmatization of the subject to a lack of clear methods of reporting such incidents, likely as few as 1 in 10 people who have these experiences ever make a report, he reasoned that the real number of actual encounters could be as high as 50,000.

For Valleé, this was just far too many to fit with the idea of extraterrestrial visitors, and is a better fit for explanations that have or more local, and potentially terrestrial, origin.

Humanoid "Aliens" Are Not Likely To Have Originated On Another Planet

His second argument is that the humanoid body structure of the alleged "aliens" is not likely to have originated on another planet—which is an idea we've talked about in previous chapters. And based on what we knew of evolution in the 1990's this is a very valid point.

Even if we admit the more recent position of some evolutionary biologists that a humanoid shape may be much more common in the universe than we might initially have thought, until we have concrete evidence that that is actually the case, we're still dealing in speculation.

And even if our understanding of evolution evolves to the point that we can more fully understand how humanoid beings could evolve independently on different planets, the actual discovery of such beings would be one of the most shocking revelations in the history of our species.

The Behavior Of These Beings Doesn't Make Sense

The third argument is that the reported behavior in thousands of abduction reports contradicts the hypothesis of genetic or scientific experimentation on humans by an advanced race.

In combing through thousands of accounts of alien abduction, Valleé recognized the common patterns that emerged in the form of the tropes that we all generally associate with the abduction phenomenon. Witnesses report being "transported into a hollow, spherical or hemispherical space and being subjected to a medical examination. This is often (but not always) followed by the taking of blood samples, various kinds of sexual interaction, and loss of time. The entire episode is frequently wiped out of conscious memory and is only retrievable under hypnosis."

And while on the surface this certainly sounds like it could be aligned with the idea that humans are being studied by a more advanced race from another planet, Valleé argues that this actually doesn't make any sense at all.

His reasoning is that any beings sufficiently advanced to create the technologically advanced craft being reported and to use those craft to travel to another planet, would be *at least* as advanced as we are in the fields of medicine and biology. And our own scientists wouldn't need to abduct and physically take genetic material from tens of thousands of different beings in order to study and understand them. So why would they?

And if, for some reason we don't understand, they did need massive amounts of human genetic material, there are sperm and egg banks all over the world. It shouldn't be a huge challenge for a super advanced race of beings to get access to those, allowing them to harvest genetic material in bulk and without all the muss and fuss of abducting thousands of individuals over several decades.

The Phenomenon Has Been Present Throughout Recorded Human History

The fourth argument is that the extension of the phenomenon throughout recorded human history demonstrates that UFOs are not a contemporary phenomenon.

As we discussed in a previous chapter, although the UFO phenomenon has been popularly regarded to have begun near the end of WWII, people have reported strange lights, objects, and beings in the sky throughout human history. Although the details of the experiences may shift and evolve throughout the centuries, any attempt to draw a line in the sand and say that any particular encounter or time period marks the definitive beginning of this phenomenon would be entirely arbitrary. After all, as Valleé points out, it's hard to find a culture that doesn't have stories of beings that come from the sky in strange vessels and occasionally abduct humans.

However, it's not just in the similarities of these stories, but in the differences between them that Valleé sees evidence that the phenomenon is more complex than the extraterrestrial hypothesis would suggest. He writes:

"In previous works I have pointed out that aerial phenomena very similar to our UFOs had been reported in the 9th century in the form of vessels in the sky, as airships in the days of Jules Verne, as ghost rockets in 1946 and as spacecraft in more recent times, as if they mimicked human expectations. Everything works as if the UFO phenomenon remained consistently one step ahead of human technology. In the last 10 years, as molecular biology has become more glamorous than electronics or even aerospace in our modern civilization, it should not be surprising to find the "Aliens" performing simulacra of genetic engineering interventions."

For Vallée, the way that the phenomenon manifests along the lines of human expectation, suggests that there is at least a component of the phenomenon that is a function of a projection of the consciousness of the observer. This is often referred to as the "Psycho-Sociological Hypothesis".

The Technology Suggests Something More Than Extraterrestrial

And finally, the fifth argument is that the apparent ability of UFOs to manipulate space and time suggests radically different and richer alternatives than the extraterrestrial hypothesis can offer.

Vallée points out that the behavior described across thousands of reported encounters indicates a level of technology that goes far beyond just the advanced propulsion systems one would need to develop for interstellar travel. These craft have been reported to exhibit the ability to appear and disappear very suddenly, to change their apparent shapes in continuous fashion and even to merge with other physical objects. All of these suggest an ability to manipulate the very fabric of spac-time. And if there are beings who have that kind of capability, he reasons that the answer to where, and even when, they come from is likely much more complicated than we can currently comprehend.

Vallée's New Hypotheses

In the conclusion of his *Five Arguments*, Valleé offers a few alternative hypotheses that he believes better explain the UFO phenomenon based on the evidence. Each is nuanced and complex, and as we separate them out into their key parts and components, we can see Valleé's startling vision begin to take shape—and with it a new world of possibilities in our exploration of the UFO phenomenon.

The Interdimensional Hypothesis

The first of these ideas is that the phenomenon could potentially be interdimensional in nature, which is referred to as the Interdimensional Hypothesis. Increasingly branches of science and physics are beginning to delve into the possibility of the existence of other universes and dimensions, and although we have no direct evidence that they actually exist, many of our scientific models depend on the idea that they do.

This potentially creates endless new possibilities when it comes to the origins of the UFO phenomenon. However, as we'll see once we dive into the individual theories, this highly speculative idea often introduces more questions than it answers.

The Ultraterrestrial Hypothesis

The second of these ideas, often referred to as the Ultraterrestrial Hypothesis, is that, although the phenomenon is alien to us, the evidence points to the possibility that it has its origin right here on Earth. This could manifest in a couple of ways.

The first would be that the UFO phenomenon is some kind of "super-nature"—something akin to the Gaia hypothesis. The Gaia hypothesis, named after the ancient Greek goddess of Earth, posits that Earth and its biological systems behave as a huge single entity. This entity has closely controlled self-regulatory systems that keep the conditions on the planet within boundaries that are favorable to life.

Introduced by chemist and inventor James E. Lovelock and biologist Lynn Margulis in the early 1970s, this new way of looking at global ecology and evolution differs from the classical picture of ecology as a biological response to a set of physical conditions.

The idea of co-evolution of biology and the physical environment where each influences the other was suggested as early as the mid-1700s, but never as strongly as with Gaia, which claims the power of biology to control the nonliving environment.

Another interpretation of the Ultraterrestrial Hypothesis that has gained popularity in recent years for reasons we'll discuss later on in this chapter, is that the UFO phenomenon is evidence of one or more races of ancient, advanced intelligent beings that have shared the planet with us, since the beginning of human history, and likely pre-dates our emergence on the planet.

And yes, I know how that sounds. Of all the ideas presented so far, for many people, this is the one that's hardest to not reject outright as being, frankly, preposterous. I get it. And I'm certainly not asking you to accept that idea as any kind of truth. All I ask is that you put it on the shelf for now.

Also, the ultraterrestrial hypothesis is a lot of fun. So my strong recommendation is that you just enjoy it.

A Control Mechanism

There's another aspect to Vallée's work, that in some ways I hesitate to introduce this early on, but it is integral enough to his research and to ufology as a whole, that I'd be remiss in not telling you about it. And that aspect is that, for Vallée, what these hypotheses have in common is that they appear to function as some kind of a control mechanism on humanity.

My hesitation in mentioning that is that this concept is far too often the gateway drug that spirals a person into fanatic belief in a pernicious and, frankly, dangerous idea that finds itself at the center of almost any conspiracy theory these days—which is that the world is being controlled by some elite cabal of satanic, probably reptilian, baby-eaters.

And while I'm not at all against the eventual exploration and reexamination of our understanding of human history and the power structures that control our government, our military, our economy, and our understanding of our place in the universe, this is an idea that we need to handle with care. There are a couple of reasons for this.

The first is that, once you decide that you're living under the influence of a control system, the invisible strings of which shape your life in ways you can't see or predict, it's very easy to lose your grasp on reality entirely. Your imagination starts to fill in the blanks and suddenly you see evidence confirming your biases everywhere and in everything. There have been plenty of examples in the news over the past several of years of normal, everyday people getting stuck in that kind of a spiral, and doing some truly horrible things from domestic terror attacks to murdering their own children as a result.

And the second reason why we need to approach this idea with caution is that, if the phenomenon is part of some sort of a control mechanism, it's likely very different from anything that we've imagined before. When we think of a "control mechanism" we're likely to think of psyops, shadowy government programs, secret societies, and men in black—but it may not be anything that literal. And we also have to acknowledge that we have no idea what the motivations behind such a system might be.

Consider, for a moment, that the UFO phenomenon could be part of a projection of our unconscious mind that is meant to somehow aid in the evolution of our species. We don't know what it was that caused sentient intelligence to awaken in the human mind—and as we'll see in future chapters—we don't really understand what caused the development of high human civilization through the discovery of things like language, astronomy, and mathematics. Could this system be guiding us somehow along this path?

Or consider the Zoo Hypothesis, which we discussed in Chapter 3. What if humans were being studied or observed in some way by a more advanced intelligence? Could the UFO phenomenon be part of a control mechanism to keep us docile and distracted in some way? Could it be concealing something about the true nature of our reality?

You can see how this idea could take you to some dark places. So we need to be careful and make sure that we're not jumping to conclusions, mistaking correlation for causation, or conveniently casting our personal villains as the villains in this story. It's an idea that we need to hold carefully, but with a light grip, as we calmly wait to gather and interpret the evidence.

Because we also can't ignore it entirely. The idea of the phenomenon as a control mechanism is an important one. It pops up again and again with different faces and with different names, but the core of the idea is always the same—that humanity is being controlled in some way by an advanced intelligence.

I think the reason that this idea is so pervasive in ufology comes down to the unpredictable and baffling nature of the phenomenon—the way that it continually shifts and evolves, always eluding us, always defying our expectations, dancing just around the edges of our understanding.

As Valleé outlined in his *Five Arguments*, when you take the evidence for the phenomenon as a whole, it is nearly impossible to piece together any kind of a coherent narrative to explain what these beings might be doing— much less the reason why. And so, it can lead one to wonder if the purpose behind these perplexing encounters is to elicit some kind of a response or to nudge us in a particular direction.

And while there is an understandable tendency to interpret this idea in a way that is fundamentally sinister—it could be that this control mechanism is actually for our benefit. It might not even be something that is controlling us from the outside, but rather a projection of our own consciousness that we don't totally understand. We simply don't know.

So let's just put this idea on the shelf for now. We'll definitely be coming back to it later.

Muddying The Waters

So this is where we'll spend some time today—by exploring the interdimensional hypothesis and the ultraterrestrial hypothesis and what they could mean. And as we move along, you'll likely notice that, although these ideas are distinct and different, they begin to overlap and intersect in interesting ways.

Which makes sense. Despite their differences, both of these hypotheses rely on the premise that our world isn't quite what we think it is, and that we are likely sharing space—at least occasionally—with one more intelligent species. And we likely have been since the beginning of human history.

Do Other Dimensions Exist?

Let's start with the "nuts and bolts" of the Interdimensional Hypothesis. If UFOs are visitors from other dimensions, we need to start with the most obvious questions:

Do other dimensions exist? And if they do, would it be possible to travel between them?

To answer this question, we need to circle back to something that we talked about in Chapter 2—quantum mechanics and the double-slit experiment.

The Double-Slit Experiment Revisited

As a quick refresher, the double-slit experiment is one of the most famous and most replicated experiments in history—which is unsurprising considering the profound and puzzling implications that it has about the very nature of our reality.

As you'll recall, light can exist either as a wave function or as a single particle called a photon. If you were to shine a light onto a board that had two slits cut into it, you would see that the light would form a pattern of lines on the wall behind the board called an interference pattern.

An interference pattern occurs when waves overlap with each other in such a way that the trough of one wave runs into the crest of another wave effectively canceling each other out. You can visualize it by thinking about dropping two stones into a still body of water a few feet apart. Each stone will form a wave that you see moving out from the stone in concentric circles. And when these ripples from the stones intersect with each other, you get an interference pattern in the water.

Waves of light work in the same way.

But let's say that instead of shining a light at the board you instead fired one photon at a time at it and there was a sensor behind the board that would show you where that photon hit. So you shoot a bunch of photons at the board, one by one, allowing enough time for the last photon to hit before you shoot the next one. What you'd find when you were finished would be that, despite the fact that the light was going through the board one photon at a time, there is still an interference pattern on the wall.

But how can that be if there were no light waves to interfere with each other in the first place? How can a single photon seemingly behave like a wave, going through both slits at once and then interfering with itself on the other side? It doesn't make sense. Based on everything we know about physics and about our lived reality in general, this should be impossible. And yet, like I said, this is one of the most replicated experiments in history—and the results are always the same.

And if that wasn't weird enough, it's this next part of the experiment that breaks everything that we know about physics and the universe and well—everything, really. Because what you find is that if you set up a sensor to detect which of the two slits a photon passes through and repeat the experiment, firing photos at the board one at a time something crazy happens—the interference pattern goes away.

The mere act of observing which of the two slits the photon goes through somehow forces it to make a decision and only go through one slit or the other. And as a result, the pattern that you get on the wall is just two straight lines.

But how can a single photon "know"—if that's even the right word to use—whether or not its path is being tracked? How does it know if we know whether it goes through one slit or the other? And what is it about our observation that changes its behavior?

What the heck is going on here?

Well, I hope you weren't hoping for me to give you an answer to that question, because although our brightest minds have been hurling themselves at the brick wall of this problem for a century now, we really haven't made any progress. We can use quantum physics to predict what will happen with stunning accuracy, but the *how* and the *why* are still a complete mystery to us.

In the words of Caltech physicist Sean Carroll, "Quantum physicists are like people with iPhones. They know how to use it, and they can do some great things with it, but if you ask them what's going on inside their iPhones they have no idea."

Of course, there are theories, but no one theory can claim consensus agreement among present-day physicists, and this debate continues to rage on. However one of the most popular interpretations, and the one that is most commonly taught in high school and college classes, is the Copenhagen Interpretation.

The Copenhagen Interpretation

The Copenhagen Interpretation was developed in the 1920s and is the shared brain-baby of famed physicist Niels Bohr and Werner Heisenberg. So how did Bohr and Heisenberg explain the double-slit experiment?

So let's go back to the first part of the experiment where a single photon is shot at the board at one time and yet still somehow creates an interference pattern on the wall. How is it that a photon can behave like a wave?

According to the Copenhagen Interpretation, this is because before the photon is measured it doesn't actually exist in a definite place or have or a particular motion. It exists everywhere and nowhere at the same time. And the waves that are interacting with each other to form the interference pattern that we see aren't actually waves of light, but rather, they are waves of probability. The photon both does and does not go down every possible path, and only materializes in a particular spot when it is forced to do so by the sensor on the wall that detects where it hits.

And so in the second half of the experiment where there is a sensor that detects which of the two slits the photon goes through, the mere act of observing what happens forces the photon to materialize before it hits the board, going through either one slit or the other, and then hitting the wall directly behind that slit.

The probability waves that were interacting to form the interference pattern before aren't there any more, because the photon has been forced to literally pick a lane and comes through the board as one single photon. So the interference pattern disappears.

According to the Copenhagen Interpretation, the photon exists in a blurry, indeterminate state as a wave function of probability. It doesn't actually have a definite position until its wave function collapses—something that only happens when it is observed.

This position is so radical and so entirely at odds with our lived experience that it took even the most devoted Copenhagen adherents a while before they began to recognize that the implication of their theory suggests that nothing is real unless it is perceived.

And as absolutely preposterous as that may sound, there is more and more evidence that it may actually be the case. In searching for answers to the mysteries of quantum mechanics, we're increasingly finding astonishing answers to age-old questions that we once thought were unanswerable.

However, for now, the Copenhagen interpretation doesn't do much for us. While it doesn't outright eliminate multiple dimensions as a possibility, this interpretation doesn't *require* other dimensions to exist—so we'll need to look elsewhere.

The Many Worlds Interpretation

The Copenhagen Interpretation's most tenacious adversary is referred to as the Many Worlds Interpretation, and was developed by Hugh Everett in 1957. According to Everett, the wave function is the true nature of reality, and therefore, it never actually collapses—it only appears to collapse because we can only see one of a potentially infinite number of outcomes.

When the observer looks to see which slit the photon goes through they see just one possible outcome, because they are only aware of one world. But in reality there are many, potentially infinite worlds, in which every possible quantum probability gets a chance to play itself out. And every world is equally real to the people living in it.

So in your world you see one outcome, but there is a world that exists for each and every possible outcome, and in each of those worlds "you" are experiencing only that outcome. And if that hurts your brain, consider that the photon isn't the only quantum system in this scenario. You, as the observer, are part of the quantum system, as well. So each of your actions and decisions plays out in this same way with every permutation of every possibility playing out somewhere out there in a world that looks a lot like this one, but isn't.

The basic premise of the Many Worlds Interpretation is that everything that can happen does happen. And I personally take issue with that, as do others in the scientific community who know a lot more about this stuff than I do. And here's why:

Grappling With Infinity

First of all, I get that the human mind's ability to truly contemplate the concept of infinity is basically nonexistent. And for me, personally, it's an idea that feels truly unapproachable. I just can't wrap my mind around it. And I always want to be mindful to not mistake the limits of my understanding for the answer.

But what the MWI requires is that there is an entire Universe erected around the infinite potential paths of one single photon—and around every other quantum system in the universe and at every imaginable scale whether we're talking about a single atomic particle, or a virus, or leaf blowing on the wind, or a human living a life, or a star going supernova.

I mean—what? I know that it's literally infinity that we're talking about here, but that still feels like—I don't know—too much?

The Free Will Problem

And beyond just grappling with infinity, there are massive and profound existential implications to this.

So imagine that you're a scientist and you're conducting the double-slit experiment. You are experiencing and observing one set of outcomes to the experiment, but every single other possible outcome is playing out in another Universe. If everything that can happen does happen, then what does that say about free will and our ability to choose?

If you make a choice between right and wrong and decide to do the right thing, does it even matter if in some other Universe you are inevitably going to do the wrong thing? And did you choose to do the right thing because it was your choice, or because this is the Universe where you *have* to make that choice.

Because with the Many World Interpretation it's not just that everything that can happen does happen, but that everything that can happen *has to* happen. And when taken to that extreme, *an infinite reality is, inherently, a deterministic reality*—and any illusion that we have of making a choice is simply the forced perspective of only having awareness of one reality at a time.

And that's unsettling. I don't think that anyone likes the idea of being just another cog in a cosmic wheel. We want to believe that we have choices and that we can shape our own destiny—or at the very least, that what we do matters.

And I personally don't think it's just the protestations of our fragile human egos that makes us feel that way. We all seem to have a deep sense that free will is intrinsic to our humanity. It's hard to fully embrace a theory that deals so dismissively with something that seems so foundational to our lived experience.

That doesn't necessarily mean that we can rule it out, but it's something to consider

Infinity And The Self

But the MWI doesn't just unravel our sense of our humanity, but of the individual self, as well. If every choice and every measurement splits off into another different version of you, each with its own unique experiences, thoughts, and sense of identity, then what even is the entity that we're calling "you" in this scenario?

Once again, this isn't necessarily a deal-breaker, but at the very least we have to recognize that the MWI interpretation, if true, obliterates our understanding of the self.

An Unprovable Hypothesis

And there are other issues with the MWI.

The problem at the heart of quantum mechanics that the MWI is trying to solve is the measurement problem—we know that measuring an event will have a direct impact on its outcome, but we don't understand the mechanism behind why or how that happens. And the MWI handles this by saying that this problem is only an illusion. The wave function never really collapses, it just looks like it does to us.

But when you think about it, that's not really an answer either. First of all, it doesn't actually solve the measurement problem, it just takes the parts that don't make sense and relegates them to other dimensions—places we, conveniently, can't access.

Many believers point to the elegance and simplicity of the MWI as self-evident proof of its reality. But elegant as it may be, you can't just take an unprovable hypothesis and assume that it's true because it appears to describe and predict reality as you perceive it. Especially when the hypothesis itself, basically by definition, predicts experimental outcomes that are fully consistent with what we observe.

I know I just said a lot of words, and if you're confused—no worries. I honestly am confusing myself. All you really need to understand is that the MWI relies on circular reasoning. We observe what we observe because it's what we can observe. There's nothing technically wrong with that statement, but it's hardly a satisfying explanation for anything.

What About The Multiverse? Is The Multiverse The Same As Many Worlds?

I want to stop here for a second to make a distinction between the Many Worlds Interpretation and the concept of the multiverse. Thanks to shows like *Rick & Morty,* these concepts have become more common in mainstream discourse, but they are often used interchangeably which can cause confusion.

As we've discussed, Many Worlds is specifically an interpretation of quantum mechanics that says that there is a Universe in which every probabilistic outcome occurs. The multiverse theory is similar but the mechanism of how the Universes come into existence is different.

For many who believe in the multiverse, a near infinite number of Universes sprang into being less than a second after the Big Bang. Each of these Universes would have its own laws of physics, its own collection of particles, its own arrangement of forces and its own values of fundamental constants—and would be unique from each other in virtually every way.

Problems With The Multiverse Theory

So what evidence do we have that the Multiverse could be real?

Well, for most advocates of this theory the best evidence for the multiverse is that we exist—complex lifeforms capable of scientific inquiry. This is because we know about our own Universe that it appears to be particularly well-suited for life. Things such as the longevity of stars, the abundance of carbon, the availability of light for photosynthesis, and the stability of complex nuclei make it possible for life as we know it to emerge.

However, if any of the starting conditions of our Universe had been even slightly different, the results could have been very different. And this is troublesome for scientists, because the chances that everything would be so perfectly balanced in our Universe that we are even able to evolve to the point to ask these questions must be vanishingly small.

And so, for many, the most logical explanation is that there must be an infinite number of Universes—the vast majority of which are likely lifeless. But inevitably, if you roll the cosmic dice an infinite number of times, you'll eventually get the conditions for life, which is how we're here.

So like the Many Worlds interpretation, what we have here is basically a blanket explanation for something that should be otherwise impossible with something that we can't access or experience, much less prove or measure.

Which isn't necessarily to say that it's wrong. But if we remember back to chapter 2, and the 2nd century AD astronomer, Ptolemy, we have a perfect example of why we need to be skeptical of these types of explanations. As you'll recall, Ptolemy was the first to challenge the idea that the heavenly bodies moved through the sky in perfectly circular orbits. They moved forward and backward across the night sky in ways that shouldn't have been possible if their orbits were truly circular.

So Ptolemy created a model that showed the planets moving in more elaborate orbits with twists and loops. His model was so accurate and predicted the positions of the celestial bodies so perfectly that it was used for the next 1400 years. But it was also completely wrong.

And so the problem with both of these theories of multiple universes is that, although they don't contradict our current models and understanding, and although they are predictive to a degree that they've allowed us to make giant strides in our technology and our understanding of the Universe, until we have a way of verifying scientifically that they even exist, it's all purely speculative. And chances are, these theories are wrong in some major way that we just can't see yet.

Multiple Universes and Multiple Dimensions

There's one more distinction that we need to make beyond just the difference between the Many Worlds Interpretation and the Multiverse—and that's between multiple universes and multiple dimensions. We often use those ideas interchangeably, but there is a difference there, as well. What we've talked about so far is multiple universes, but multiple dimensions would work differently.

Think of it this way—we are 3rd dimensional beings that exist in the 4th dimension, which is time. To begin to understand how wildly different a different dimension would be from a different Universe, just try to conceive of what a 4th dimensional being living in the 5th dimension might look like.

Having trouble? Yeah, my brain doesn't work that way either. But for the purposes of this conversation, it's not important that you be able to visualize or understand dimensions—only that you know from a high level what they are.

String Theory And The 11th Dimension

And no conversation about dimensions would be complete without discussing String Theory. String theory proposes that the fundamental constituents of the universe are one-dimensional "strings" rather than point-like particles. What we perceive as particles are actually vibrations in loops of string, each with its own characteristic frequency.

I'll be real and tell you that I don't totally get String Theory either, but right now we don't need to. There are basically two things to know about String Theory that are worth mentioning here.

The first is that, unlike the universes in Many Worlds Interpretation or the Multiverse Theory, the dimensions proposed by String Theory aren't infinite. In fact, there are specifically 11 dimensions. This has to do with how the dimensions fold into each other like nesting dolls.

However, within these 11 dimensions String Theory can reasonably predict a multiverse populated by 10^{500} different universes. And while that's not technically an infinite number, it might as well be. After all, 10^{82} is the number of atoms in the Universe—so 10^{500} is just incomprehensibly large. That would mean that there exist more Universes in the multiverse than there are atoms in our own Universe.

And like the Many Worlds Interpretation and the Multiverse Theory, String Theory is still entirely hypothetical. Thus far we have no way to verify if it's true or not.

The Verdict On The Interdimensional Hypothesis

That was a lot of information. So after all of that, where does the Interdimensional Hypothesis shake out? Well, like anything else having to do with the UFO phenomenon, the answer is complicated.

As we've established, we don't have any direct evidence that other Universes exist. We know that other dimensions exist, at least so far as we know that we are 3rd dimensional beings that exist in the 4th dimension, but exactly how dimensions work and whether or not we could even perceive beings that came from another dimension is all highly speculative.

And this can be extremely unsatisfying for anyone looking for a solid, scientific footing from which to assess this hypothesis. Each of these theories relies on the concept of infinity to explain phenomena that would be otherwise unexplainable, and all of them are ultimately unprovable because we can't perceive or access these other universes and dimensions.

However, it is worth considering the fact that some of the best and brightest minds of the last century have spent their careers working on some of the biggest and most baffling questions that science has to ask—and many of them have come up with some version of the same idea to answer it. And those models, though unprovable, are still predictive enough to allow us to make groundbreaking breakthroughs in technology ranging from LED lights to quantum computers.

So maybe there's something to this after all?

Ptolemy's model of the planetary orbits wasn't correct, but it was a vast improvement over the models that came before it. And it more than served its purpose for 1400 years, until the next leap in our understanding allowed us to discover their true orbits.

And so it will likely be with our understanding of parallel worlds and dimensions when we finally take that next step in our evolution. Our current theories will likely prove to be much the same as Ptolemy's model of the cosmos—partly true, partly false, and with some vital truth still yet unknown to us, suddenly revealed.

Could Paranormal Phenomena Be Interdimensional?

One more idea I want to touch on here before we move on is what the Interdimensional Hypothesis suggests, not just about the UFO phenomenon, but about the paranormal in general.

The word paranormal simply refers to things that are beyond the scope of normal scientific understanding. And while the scientifically minded among us can be quick to dismiss the paranormal as ridiculous, the reality is that this realm beyond our understanding is where many of the brightest scientific minds are spending most of their time—and from whence most of our most startling breakthroughs are being made.

We like to think that science "sees all" and "knows all", but science is just a method of inquiry that allows us to slowly and incrementally move back the veil that keeps the truth behind the mysteries of the universe shrouded from our sight, revealing a deeper and deeper understanding of what is.

And while it's impossible to say what lies just beyond that veil, many scientists believe that what is there could very well be infinite in nature.

And the thing about an infinite multiverse is that if it is possible for beings to travel between different worlds and dimensions, then someone somewhere—and likely very many someones—would definitely figure out how to do so. And in that scenario, it's not so much a matter of *if*, but *when* and *how often* humans would encounter them.

And these beings, whatever and whoever they might be, could look a lot like us—potentially exactly like us—or they could come from a Universe that is so different that we wouldn't even necessarily recognize them as lifeforms at all.

So could UFOs potentially be visitors from another Universe or another dimension—absolutely. But if that's possible, what else might be coming through the veil that defies our understanding?

The Ultraterrestrial Hypothesis

And now we arrive at perhaps the strangest of the possible explanations for the UFO phenomenon—the Ultraterrestrial Hypothesis.

Lue Elizondo & *Chains Of The Sea*

One of the main reasons that the Ultraterrestrial Hypothesis has been getting so much attention lately has been as the result of comments made by the former head of AATIP himself, Lue Elizondo.

When listening to Lue speak, something that becomes immediately obvious is the care with which he chooses his words—and he has to. As a former member of the U.S. military and intelligence apparatus, he still maintains his security clearance and has an iron-clad NDA. This creates a curious situation where all the things that people most want to talk to him about are often the very things that he can't talk about.

And yet, Elizondo is still adept at giving engaging and thought-provoking interviews on the UFO phenomenon by reframing questions and his responses in ways that allow him to speak broadly and speculatively about the subject without technically giving away any state secrets. He has said openly that he leaves clues and breadcrumbs for people to follow, and as a result, every time he does an interview, there's a large segment of the UFO community that will spend weeks dissecting and analyzing everything he says looking for clues.

And there's one statement in particular that has gotten a lot of attention and has caused many people to give more consideration to the Ultraterrestrial Hypothesis. During an interview, Lue was asked if there were any fictional works that people could read that would give them an idea of what the phenomenon is, and his answer was surprising.

Lue recommended a little known book of short stories called *Chains of the Sea*—and specifically the second of the three stories in the trilogy by the same name. I haven't actually gotten an opportunity to read the book—it's so rare at this point that copies go for $500—but having read several summaries and excerpts from people who have read it, here is the general gist:

Alien ships arrive on Earth, landing in Delaware, Ohio, Colorado, and Venezuela. Humans, true to form, try—unsuccessfully—to fight the aliens and then desperately try to cover up the landings. The aliens, however, have very little interest in humans and basically ignore them, choosing instead to communicate with AI that has been created by the humans.

The AI has evolved to the point where it is smarter than humans, and it has figured out a way to hide communications from them through back channels. In these talks, the aliens reveal to the AI that Earth isn't ruled by humans or AI, but by previously unknown races of non-human intelligent beings with whom we share space, but that can only be seen sometimes by some people.

Interesting, right?

Now, it's important to pause here to say that Lue has said multiple times that he didn't mean to imply that anything about the plot of this story is literally true. Instead, he says that the story is useful because it helps people think differently about the phenomenon.

On October 15, 2020 he tweeted:

"[Chains of the Sea] has little to do [with] the sea [and] more to do [with] the possibility of

an ultraterrestrial universe. My recommendation of the book (2nd of 3 short stories) is the way in which the story (fiction) is told. Forces the reader to consider other possibilities about the nature of the phenomenon."

So while *Chains Of The Sea* may not be literally true, it does open up a vast and exhilarating new world of possibilities. And hearing someone with so much privileged, insider knowledge make such a strong case for what in his words is an "ultraterrestrial universe"—it's hard not to let your imagination run wild.

So let's take a little time to play with this idea. As you might imagine, this topic is sprawling and more than a little bizarre. We'll only just be able to scratch the surface in this chapter, but we can at least explore the high points, and from there take in some of the more exotic vistas it has to offer.

Could There Be Species On Earth We Don't Know About?

So to start with, we need to examine the, frankly startling, idea at the heart of the Ultraterrestrial Hypothesis which is that we could potentially be sharing our planet with one or more intelligent, non-human species without being aware of it. Is that even possible?

Are Some Cryptids Real?

Let's take the case of cryptids. If you aren't familiar with that term, a cryptid is an animal that some people claim to exist, but whose existence has never been definitely proven. Popular examples that come to mind include BigFoot or the Loch Ness Monster. However, contrary to popular belief, cryptids don't have to be supernatural, mythical, or even all that strange—just unproven.

There are actually several species that used to be cryptids, but were eventually discovered to be real—and some of them are pretty freakin' weird.

Cryptids That Turned Out To Be Real

1. **Komodo Dragon**
   One classic example of a cryptid that turned out to be real is the

Komodo Dragon. For centuries, stories of enormous lizards on the island of Komodo in Indonesia were laughed at by any respectable scientist in the Western world. That is until 1910 when the first specimen was caught and killed, shocking the world. A pair of Komodo dragons that was kept at the Bronx Zoo even inspired Merian C. Cooper to write the 1933 classic *King Kong*.

2. **Platypus**
   Not quite as terrifying, but in many ways just as unlikely, is the platypus. Looking like a bad photoshop of a duck, otter, and a beaver all mixed together, most people didn't believe that the platypus could exist—and it wasn't until the 19th century that it was definitively proven to be real. Scientists were so convinced that this absurd creature couldn't exist that the first few pelts that were produced were widely regarded to be hoaxes.

3. **Gorilla**
   Starting in the 5th century A.D., European explorers reported seeing hairy, humanlike monsters in Africa. However, it wasn't until 1847, when Thomas Savage found gorilla bones in Libera that gorillas were first recognized as a real species, and even then it took another decade for one to be captured and analyzed. Famously private, one remote species of mountain gorilla wasn't even discovered until 1902.

4. **Giant Squid**
   Though they've been reported by sailors for millennia, giant squids were considered to be a myth. However, that changed when scientists first took images of a giant squid in 2004, and then was finally put to rest in 2006 when a 24-foot long female was caught alive by Japan's National Science Museum.

These are just a few examples of cryptids that turned out to not be cryptids.

And before we move on, I think it's important to point out that in most of these cases, cryptids probably weren't cryptids to the people who actually lived there. For example the people of Liberia undoubtedly knew about gorillas before European scientists came and confirmed their existence for themselves.

There is a tendency, especially for those of us who grew up with a Eurocentric worldview, to apply the blueprint of what is familiar to us to everything, which causes these kinds of errors. And it's a good reminder that it's impossible to make a genuine scientific inquiry into anything if you've already decided ahead of time what is possible and what is impossible.

Where Could Cryptids and Ultraterrestrial Species Hide?

So we know that some animals that we once considered to be cryptids turned out to be real. So it stands to reason that it's at least possible that there could be a non-human intelligent species somewhere on the planet that we haven't encountered yet. And if they are as technologically advanced as the UFO phenomenon would suggest, they'd theoretically have a much easier time hiding from humans than gorillas or even giant squids.

But that begs the question—where could they be hiding?

Other Universes/Dimensions

One of the most obvious places is somewhere we've already talked about—which is in other Universes or dimensions. This is where the Interdimensional Hypothesis and the Ultraterrestrial Hypothesis start to blend together. Basically, it could be that we occupy more-or-less the same place in space and time as other beings, but that we usually can't see or interact with them, except for under special circumstances.

In terms of how that would actually work—we have no idea. But as we've already explored, the potential here is literally infinite.

Underwater

However, these beings wouldn't necessarily have to hide out in another dimension in order to avoid our detection. There are still plenty of places on Earth that we can't access and about which we have virtually no information—and almost all of them are underwater.

Our Unexplored Oceans

Our oceans account for 70% of the surface of the Earth, and yet we've only explored about 5% of them—and almost all of that is very near the surface. Ironically we know more about Saturn than we do about the deep oceans on our own planet.

It's the very vastness of the oceans that creates a challenge in exploring them. Saturn is literally a billion miles away from the Earth, but you can see it in the sky with your naked eye and its rings can be detected with even the smallest telescope. However, once you go deeper than 200 m (or about 650 feet) in the ocean, the sunlight can no longer penetrate and it gets very dark, very fast.

And when you consider that the oceans can be up to seven miles deep, it immediately becomes clear that a staggering proportion of what we would consider to be the livable space on this planet is completely uncharted and unknown to us. It's hard not to consider what might be hiding down there, but at the very least there are likely countless species and ecosystems that exist on this planet of which we have no awareness.

UFOs & Water

This underwater possibility is especially interesting to consider because there is an undeniable relationship between the UFO phenomenon and water. Lue Elizondo himself has brought up in multiple interviews and on his History channel show, Unidentified, that UFOs show one other seeming affinity besides nuclear facilities—and that they are very often found around water.

Not by coincidence, all of the declassified Navy videos of UFOs happened over water. And it's very interesting to note that these declassified videos and the ultimate confirmation about the existence of UAPs has come from the Navy—while the Air Force has remained conspicuously quiet. That's a conversation for another chapter, but it further underlines the connection between the UFO phenomenon and water.

And it's not just ufologists and researchers here in the States that have noticed this seeming affinity. In 2009, Russia declassified a bunch of its Cold War era documents reporting the UFO phenomenon and they had reached the same conclusions.

One document reads, "50 percent of UFO encounters are connected with oceans. Fifteen [percent] more—with lakes. So UFOs tend to stick to the water."

The documents contained several reports of underwater encounters with USOs (or unidentified submersible objects) including incidents with multiple objects moving in formation under the water at speeds of up to 230 knots (or 400 kph)—something that should be physically impossible given the water resistance.

The Mystery Of Lake Baikal

And it's not just in the oceans where these encounters are happening. Large lakes have also been known to be epicenters of UFO activity. One example is Lake Baikal in Russia. Lake Baikal is the deepest lake in the world and contains more than 20% of the world's freshwater. Reaching a mile deep in places and with temperatures dipping down to a frigid -19 degrees Celsius (or -2 degrees Fahrenheit), not much is known about Baikal's icy depths.

Russia's declassified document dump included multiple reports of mysterious lights and objects in skies above Lake Baikal, as well as reports of them coming in and out of the water itself—something that locals and fishermen have been reporting for years.

According to one of these declassified documents, in 1982 seven Russian Navy divers were conducting a research mission in Lake Baikal. The divers were at a depth of 50 meters when suddenly they realized that they were being watched by unknown beings under the water. These beings were reportedly anthropomorphic, but much larger than humans at 9 feet tall. They were reportedly wearing some sort of super thin silvery suit, and while they appeared to have some sort of a helmet over their heads, there was no evidence of scuba gear or any similar breathing apparatus.

The Russian divers then allegedly tried to capture one of these beings, who reportedly responded with some sort of a powerful, unseen force that propelled all of the divers rapidly to the surface. Three divers died from the bends as a result of this rapid ascent, and the other 4 were severely injured.

Wild, right? Now, granted, I have no idea what to make of this account and there's no real way to verify any of it. But still, the fact that this report was apparently made and documented by the Russian Navy at a time when the stigma around the UFO phenomenon was at its highest is more than a little bit interesting.

Underground

Another possible hiding place for a non-human intelligent species would be underground. The reality is that, in many ways, we know even less about what lies beneath our feet than we do about our deepest oceans.

So how far underground would it even be possible for life to exist?

That's not an easy question to answer. Geo-microbiologists have found bacteria living in rock two miles beneath the Earth's surface. So there is evidence that subterranean life is possible. However, the organisms we find at that depth are extremophiles, enduring conditions that most living organisms couldn't withstand. And it's almost certainly not bacteria that's creating and piloting UFOs, so that's not super helpful.

Caves would be a better candidate for concealing complex lifeforms, but our information about caves is somewhat limited, as well. We do know that there are a lot of them. In the United States alone there are 17,000 caves, with likely thousands more still undiscovered. And some of these caves go deep underground—the deepest that we know of is 2,212 meters (or 7,257 feet) deep.

An advanced civilization could conceivably even expand upon existing caves, creating deep caverns and even cities underground. This might seem far-fetched, but building and living in cities underground is something that humans have done at times in our history, as well. And while underground life might sound dreary, if you're looking for protection from the elements or your enemies, it's very effective.

One famous example is Derinkuyu, a massive underground city that is actually one of a complex of underground structures in Cappadocia, Turkey. The city, which is estimated to date back to the 8th century AD—although there is debate that it could be much older—spanned more than 4 million square feet and could house up to 20,000 people. Even by today's standards it's an architectural marvel.

So while an underground civilization may sound outlandish, it's at least possible in the technical sense, making this strange theory one that we still can't rule out.

Hollow Earth

Since we're here and we're talking about ultraterrestrials hiding underground, we absolutely have to discuss one of the wildest theories out there—and a personal favorite of mine—the Hollow Earth Theory. This is the idea that our planet (and some extend this idea to include all planets) is, in fact, hollow and that it is teaming with life, and potentially hidden civilizations.

I love the Hollow Earth Theory. To be clear, I don't believe it's true. The physics of a hollow planet just simply does not check out, and the little hard evidence that is offered up is highly "sus", as the kids say.

But come on—this one is just fun.

If you're unfamiliar with the idea of the Hollow Earth, there are many different permutations of this story, but the basic idea is that, you guessed it, the Earth is hollow. Many adherents believe that there are giant openings to the interior of the planet at the north and the south poles, and that one or more highly advanced species have made their homes in cities deep underground.

Generally, everything in the Hollow Earth is like the surface of the planet, but better. The civilizations found there are peaceful and live in harmony with their environment. The environment is unpolluted and better suited to life, allowing for the evolution of plants, animals, and beings that are bigger and far more exotic than one would find on the surface.

As I mentioned, the physics of it all is baffling. How a hollow planet wouldn't collapse in on itself is a real mystery. And when you add in the fact that most versions of the Hollow Earth include an interior sun suspended in the middle of the planet, it doesn't just push the lines of credulity, it shatters them.

Still, this is an idea with surprising staying power. It was first introduced in the 1700s by physicist Edmond Halley and despite centuries of debunking, it continues to spring up anew with a few details changed, but with the same fantastical belief still firmly at its center.

I think one of the reasons that I like Hollow Earth so much is that it is literally and figuratively the opposite of the Flat Earth theory. Literally they're opposites in the sense that, instead of a flat Earth, this model turns the Earth in on itself creating another secret world inside.

But also, I find them to be spiritual opposites, as well. The Flat Earth is less than what we thought it was, and somehow far more sinister. The

Hollow Earth is more than we thought it was, and is full of fantastical possibilities and the promise of a better world. True or not, I know which one I'd rather spend my time on.

And, if I'm being honest, that's where I shake out on a lot of this stuff. It is inherently human to want to know the unknowable—to approach the unapproachable. But when we wander out beyond the realms of current human understanding, we find ourselves unmoored. Like Ptolemy, we have a model of the Universe that fits, but doesn't quite fit—that looks true, but that still hides some deeper mystery awaiting the day when our perspective swells to a size that can contain it and we can finally see it for what it is.

And in this space of wonder and imagination, where we walk right up to the veil without yet having the means to move it aside to see what lies beyond, the ideas that we cultivate and entertain are, by their very nature, more a reflection of who we are than of what we are hoping to reveal about the nature of reality.

Which is why, right or wrong, you'll always find me on the side of the dreamers and the aggressively non-cynical.

And that's where we'll pick this up next week, with an introduction to the cast of characters at the heart of the growing disclosure movement through the lens of what is perhaps the most surprising and least cynical stories ever told—and one with the unlikeliest of heroes.

6 Mr. Delonge Goes To Washington [Part 1]

We've spent the past five chapters talking through the possible answers to the most obvious questions raised by the Pentagon's stunning admission that UFOs are real—which are, who's making them? Where are they coming from? If these craft are intelligently designed and controlled, what is the source of the intelligence behind them?

And, as we've seen, there are no easy answers. The UFO phenomenon is hard to grasp with our current understanding. The tighter we try to hold onto it, try to make it conform to our preconceived notions and expectations, the more easily it slips through our fingers—and as it smashes on the floor, glittering shards of possibility skittering in all directions, we begin to see that this phenomenon has much more profound implications than just the potential of visitors from another world.

And, to be honest, there has been no aspect of my trip down the UFO rabbit hole that has surprised, confounded, terrified—but also, frankly, delighted me—quite like the story that we're about to cover today. This story has everything: twists and turns, rock stars, clandestine meetings, government secrets, good guys and bad guys, angels and demons, UFOs, of course, and an unlikely hero who through sheer grit and determination accomplishes the impossible.

And much like the UFO phenomenon, this story has been hidden in plain sight—which doesn't mean that it was necessarily easy to find. As we dive into this story, you might find yourself wondering—how the heck have I never heard all of this? And you're not alone.

While the mainstream media has taken sporadic and frustratingly shallow interest in this story, it's been largely overlooked. But behind-the-scenes, at the highest levels of our government, the events that we're going to discuss have forced a seismic change in how the UFO phenomenon is being addressed both internally and to the public..

And right at the center of this story is pop-punk icon, and our unlikely hero, Tom Delonge. Yes, that Tom Delonge from Blink-182. And we'll get into that whole story in just a minute.

But first I think it's important to get a sense of who Tom Delonge is. And not just because I will forever go to the mat for early 2000s pop-punk as not just an artform, but a lifestyle. But because of the way that his life and career brought him to a surprising turning point, making him one of the most—if not *the* most—important figures in the history of the UFO disclosure movement.

Tom Delonge: Early Life

Thomas Matthew Delonge was born on December 13, 1975 in Poway, California, a small city of about 50,000 people just north of San Diego. The son of an oil company executive and a mortgage broker, Delonge knew that he didn't want to, in his words, "grow up and get a job that [he] hates". School wasn't of much interest to him, so unsurprisingly, he wasn't a great student. However, he had an endless enthusiasm for his passions outside the classroom, saying, "I knew exactly how hard I had to work in school. As long as I got that C, I wouldn't try one minute extra to get a B. I just cared about skateboarding and music".

Tom first picked up a skateboard in 3rd grade and it quickly became an obsession. He describes this time of his life saying, "I lived, ate, and breathed skateboarding. All I did all day long was skateboard. It was all I cared about." After receiving a beat up guitar as a gift from two friends for Christmas in the sixth grade, Tom had dual obsessions to occupy his time.

But planted alongside these seeds of the cultural icon that he would become, was another life-defining passion, one that would eventually grow to eclipse them all. In junior high, while being forced to spend quiet time in the library, he searched the stacks for something to read that didn't fall into the category of what he called "boring shit". It was there that he found a book with pictures of the Loch Ness Monster and a UFO on the front. Tom devoured the book, utterly enthralled by the strange ideas inside, sparking in him a lifelong obsession with the topic of UFOs and the paranormal.

Tom Delonge: Formation of Blink-182 & Career

In 1992, after being kicked out of Poway High School halfway through his junior year for showing up drunk to a basketball game, Tom found himself as the new kid at Rancho Bernardo High School. He signed up to play in a Battle of the Bands organized by the school where he played an original song poignantly titled, "Who's Gonna Shave Your Back Tonight?" to a packed auditorium. It was there that he began to make the connections that would lead to the formation of Blink-182 alongside bassist and co-vocalist, Mark Hoppus, and drummer, Scott Raynor.

The three quickly became inseparable, spending hours in Raynor's bedroom writing and playing music, pausing only for punk concerts, skateboarding, and practical jokes. A teenage boy's dream, right?

But Tom wanted more. They released their first album, *Flyswatter*—a combination of original punk songs and covers—and soon after Delonge began calling around to venues all around San Diego trying to get stage time for the fledgling band wherever he could. Showing an early tenacity that would drive him throughout his career, he even called local high schools, telling them that Blink-182 was a "motivational band with a strong anti-drug message" in order to book assemblies and school events.

Their persistence paid off and by 1995, the band was touring constantly throughout the United States, Canada, and Australia. There was no WiFi back then, if you can imagine such a thing, and so Tom passed the endless hours on the tour bus traveling between shows devouring stacks of books on UFOs and the paranormal. It was during this time that he first read the works of some of ufology's greatest founding fathers including: Jacques Valleé, Stanton Friedman, and John Keel.

By 1996, the band had gained enough buzz to spark a bidding war between Interscope, MCA and Epitaph. The band ultimately signed a record deal with MCA, who offered them the most artistic freedom. However, disagreements over this deal, along with Raynor's substance abuse issues, led to the band parting ways with the drummer in 1998. He was quickly replaced with Travis Barker, the drummer of another band they were touring with, The Aquabats, after he filled in for Raynor, famously learning the band's entire set list in the 20 min before the show.

In 1999, Blink-182 released their seminal album *Enema of the State*, which immediately rocketed them to crossover stardom, dominating not just the rock charts but the Billboard Hot 100, as well. The sound that the band had developed, while heavily influenced by their punk rock idols like NOFX, The Descendents, and Bad Religion, was infused with the sunny optimism of Southern California, and radio-friendly hooks. And unlike punk rockers of the past, Blink-182 never shied away from their commercial appeal. They even had a contract with Billabong (as literally any picture of the band from the 1998-2002 can attest), making them the first band ever to be sponsored by a clothing brand.

Even as the stadiums began to fill and Blink-182 began its meteoric rise to fame, leading the charge in a new wave of pop-punk bands, Delonge's interest in and fixation on the topic of UFOs never wavered. In interviews, his bandmates recount endless hours of listening to Tom ramble on about aliens and government conspiracies, things for which he had both an encyclopedic knowledge and a seemingly inexhaustible enthusiasm. On breaks in between cities he was known to organize trips to remote locations to look for UFOs, BigFoot, and other manifestations of high strangeness.

And through this time, Delonge remained remarkably grounded. He seemed well aware of the "flash in the pan" nature of fame and despite Blink's meteoric rise, he has had multiple successful business ventures on the side throughout his time in the spotlight.

Delonge established a holding company, *Really Likable People*, in 1998. Under this holding company he developed multiple clothing brands, a shoe brand, and even a technology and design firm that has handled the official websites and fan clubs for a range of popular artists, including the White Stripes, Pearl Jam, and Kanye West.

However, fame was more than just a "flash in the pan" for Tom and his bandmates. With their 2001 release, and first album to debut at #1, *Take Off Your Pants and Jacket*, Blink-182 solidified themselves as both Billboard Chart mainstays and cultural icons.

Delonge at times seemed to resent being boxed in artistically by being part of one of the biggest and most commercially successful bands on the planet. Feeling an "itch to do something where he didn't feel locked in to what Blink was", in 2002 he released a new album under a new name *Boxcar Racer*. Although he wrote the songs himself, for convenience, he asked Travis Barker to record the drum tracks for the album. This caused tension between Tom and Mark Hoppus, who was not included on the project and marked the beginning of a rift between the two that only widened throughout the coming years.

By 2004, tensions boiled over and Blink-182 broke up. Reeling from this turn of events, Delonge looked to reinvent himself. He went on the road with John Kerry's presidential campaign and discovered a new side of himself—one that was driven by the desire to bring about positive change in the world.

In the wake of this sweeping internal change, Delonge formed the band Angels & Airwaves, a project that he saw as being much more than just a band, but "an art project [that approaches] larger human themes and tackles them in different mediums". He started talking about films and multimedia projects that he predicted would change and influence the youth culture, creating a revolution through art.

So by the late 2000s, we see this new version of Tom Delonge begin to take form—one who is equal parts businessman and artist, and one who is passionate about finding ways to use his considerable talents and resources in both arenas as a catalyst for change.

Blink-182 got back together in 2008 with more albums and tours to follow, but the tensions between the band quickly resurfaced despite their success, particularly between Hoppus and Delonge. The duo had a bit of Lennon and McCartney vibe, with Tom, despite his class clown public persona, often playing the mercurial and restless Lennon to Mark's sunshiney and commercial McCartney. But in the end it was UFOs that would be their Yoko Ono. (I promised myself I wasn't going to leave that line in, but the heart wants what it wants.)

2014: Tom Delonge's First Encounter With The Phenomenon

In 2014, Delonge had an experience that would change the course of his life forever. Here is the story as he recounts it.

Delonge had been in contact with a well-known ufologist who believed that UFOs communicated through consciousness and who claimed to be able to make UFOs appear through a set of meditation protocols. I find it interesting that he's never come right out and identified this individual, because to anyone who is even casually plugged into the UFO community this can really only be one person—Dr. Steven Greer.

There are some theories as to why this might be that we'll get into in the next chapter, but for now you just need to know that Dr. Greer has been a prominent, albeit controversial, member of the UFO community and vocal advocate for disclosure since the early 90s. He's most well-known for having developed the CE5 (which stands for "Close Encounters of the 5th kind") protocols, which is basically a series of meditations that you can do that he claims can be used to summon UFOs.

So in 2014, Delonge and this "unknown ufologist" go camping in Death Valley with a few other people. Their plan is to do these meditation protocols and try to make contact with a UFO. They do this for a couple hours without success and eventually decide it was time to go to sleep.

Delonge claims that he was later woken up in the middle of the night to the sound of hundreds of voices outside of his tent. He couldn't quite make out what they were saying, but he wasn't able to make himself move to see what was going on and ended up quickly falling back asleep.

When he woke up, he discovered that one of the other members of his group had had the exact same bizarre experience, leading Delonge to believe that his experience was real. This baffling encounter with the phenomenon shook him to his core, leading him to reevaluate his purpose and direction in life.

Tom Delonge Leaves Blink-182

In 2015, Delonge abruptly quit Blink-182, telling his bandmates through his manager that he wanted to focus on his "non-musical" endeavors. Delonge later claimed that he never actually left the band, saying he "Never planned on quitting, [I] just find it hard as hell to commit." Whatever his intention, the indefinite nature of his hiatus and deteriorating relationships within the band led Hoppus and Barker to replace Delonge with Alkaline Trio's Matt Skiba with whom they continued to tour and record.

Despite any hard feelings that Delonge may have had over this turn of events, it's clear that he more than had his hands full with his new endeavors. It was more clear to him than ever that the phenomenon was real, and he was determined to play a role in bringing this issue to the public awareness. And always a man of action, Tom Delonge had a plan.

To The Stars

In 2015, shortly after leaving Blink-182, Tom Delonge founded an entertainment company called To The Stars. He said in later interviews that a major inspiration for this endeavor was the movie *13 Days*, a 2000 film dramatizing the events of the Cuban Missile Crisis. Delonge recounts that after seeing the film, he found himself suddenly intensely interested in the topic in a way that he never had been in school, wanting to read everything he could get his hands on.

Seeing the potential that entertainment has to help people engage with and understand complex topics they might otherwise ignore, Delonge formed To The Stars to do just that. In his opinion, UFOs had never been portrayed correctly in the media, and with TTS he saw the opportunity to slowly disclose the nature of the phenomenon to the public through entertainment and media to include films, documentaries, tv shows, books, and more.

However, Delonge also recognized that if he was going to be truly successful in helping people come to grips with the true nature of the phenomenon that he would need some help in understanding it himself. He'd read about every book on the subject that existed and certainly had his own opinions, but without some kind of privileged, insider knowledge, he had no way of knowing for sure exactly what was true and what wasn't.

If he was going to do this right, Tom Delonge needed advisors.

And according to Delonge, a stunning and wildly unlikely series of events unfolded over the course of the next two years that would bring him those advisors from the very highest levels of our government, military, and intelligence agencies.

And this is where the story really starts to get strange. Over those next two years in 2016 and 2017, Delonge proceeded to give a series of increasingly bizarre interviews in which he made shocking claims pointing to anonymous, mysterious sources within the government whom he could not identify.

These interviews can be as frustrating to listen to as they are fascinating. They're frustrating because, although it's clear that Delonge wants to share every detail of what he's been experiencing, he is trying to do so while not revealing any potentially identifying information about his sources. This leads to a lot of incomprehensible stories about going to a "certain city" and being taken to a "certain location" to speak with "certain individuals". It's not easy to parse, and basically all of it—at least at the time—was impossible to verify.

However, these interviews are also fascinating, because in them you find a frenetic and entirely unvarnished version of Delonge. He talks too fast, he says too much, and you can almost hear his frustration at our inability to keep up. And while he speaks cryptically about the sources of his information, he speaks specifically and with the conviction of a man whose deepest beliefs have been proven to him beyond a shadow of a doubt about things that sound shocking, bizarre, and to be honest, downright insane.

His claims fall into two main categories that we will address one at a time. The first is how he assembled his group of high-level advisors, and the second, which we'll address in-depth in part 2, is what those advisors have told him about the nature of the phenomenon.

What follows is an inevitably imperfect accounting of what Tom has said about how he assembled his advisors, pieced together from dozens of hours of interviews from Tom Delonge himself, many of which occurred in 2016 and 2017 as these events were still unfolding.

Tom Delonge Assembles His Advisors

So going back to 2015, Tom recognizes that for TTS to be successful in its mission, he's going to need advisors with insider knowledge of the nature of the phenomenon. And almost as if by magic, a surprising opportunity presented itself out of the blue—leading to one of the most insane and unlikely series of events, honestly maybe ever.

The Aerospace Executive

As he tells it, Delonge was asked to introduce the lead aerospace executive at an open house for families of a top defense aerospace contracting company. Fortuitously, this was the first time that this organization had ever had an open house of this kind where families and other members of the public were invited to attend.

There's no way to know for sure, but many suspect that Delonge is referring to Lockheed Martin, and more specifically to their notorious and secretive special projects division, Skunk Works. As you may recall from Chapter 2, the Skunk Works division of Lockheed Martin was founded in 1943, during WWII, when the U.S. government needed to quickly develop the country's first jet fighter to compete with the new German jets that were appearing in the skies over Europe. Since then, Skunk Works has been a highly favored government contractor taking on the development of some of our most highly classified experimental aircraft, which are tested and developed at Area 51 and at Groom Lake.

Delonge, knowing exactly who this executive was and the nature of his work, seized this opportunity and said that he would agree to introduce this aerospace executive if he could get five minutes alone with him afterward. They agreed and soon after introducing this executive at the open house, Delonge found himself alone in a room with him

Not knowing how his pitch would be received, Delonge nevertheless dove right into it. Intentionally avoiding the word UFO for fear of getting immediately shut down, he instead presented his idea as a way for the youth to become less cynical about government secrecy and the military industrial complex.

The meeting was brief and the executive didn't say much, but he set up a meeting with Delonge to discuss the matter further.

According to Delonge, this meeting took place at an undisclosed, highly secure location where he had to pass through four layers of security—past heavily armed guards and through hallways flooded with white noise to prevent conversations from being overheard—into a small, windowless room deep within the building.

In this room, he met with the first aerospace executive that he had spoken to along with other high-ranking aerospace engineers. Delonge began his pitch again as before, carefully avoiding the UFO angle and focusing on the benefit to the government and its contractors.

However, one of the executives had been Googling Tom and was aware of his belief in UFOs and his penchant for conspiracy theories. This person asked him outright, "So what do you plan to do with all of this conspiracy stuff that you're into?"

Delonge did his best to dodge the question, but despite his attempts to gloss over it, the executives steered the conversation back to the subject of UFOs, finally saying, "We cannot be associated whatsoever with anything that this has this topic associated with it, specifically because there has never been any evidence whatsoever that this even exists."

At this point, Delonge thought that he had hit a brick wall, but he dug in his heels and responded, "If Edgar Allen Mitchell, the sixth man to walk on the moon, is out telling every kid in the world that this topic is real, then we have a problem."

This seemed to give them pause, and Delonge jumped on the opportunity and boldly asked to speak to the lead aerospace executive alone. Sensing that this was his moment, he said to the executive:

"I want you to understand something. I understand the national security implications of what I'm about to say. I am not naive to the topic. I think if you hear me out you'll see that there's merit in what I'm about to propose. Over the past 30 years there's been a program to indoctrinate people to the idea that this might be real, but the problem is that all the young adults of the world, they use the internet, they have iPhones, they talk to each other much quicker than people ever have. So this program that everyone's been following from the 50's is far outdated. It's antiquated. People have surpassed it and now they don't trust you guys. Now they don't like you guys. Now they graduate from MIT and they want to work for Elon Musk and they don't want to work here. Help me help you guys."

The lead executive seemed swayed by this and after some more conversation he eventually admitted that "Yes, UFOs are real" and that he thought that Tom's plan of slow disclosure could actually work in revealing the reality and nature of the phenomenon to the public.

Following this conversation, the executive then connected Tom to someone whom he referred to as "The General".

The General

After this introduction, Delonge begins to correspond with The General. And after a lot of back-and-forth over email, they eventually meet secretly in a random airport that they both flew into specifically for this meeting.

Tom sits down at the table with the guy and without any prelude or small talk The General says to him,

"It was the Cold War and we found a lifeform, and every single day we lived under the threat of nuclear war, every single day we really believed that nuclear war could happen at any moment. And somewhere in those years we found a lifeform. And everything that we did and every decision that we made with that lifeform was because of the consciousness at that time."

He went on to tell Delonge that during the Cold War, UFOs would routinely interfere with US and Soviet nuclear weapons in what looked very much like an attempt to instigate nuclear war—and that they were very nearly successful. He said emphatically that there were heroes in Russia who did not fire back when UFOs manipulated radar systems to appear as though Americans had launched a first strike.

As they concluded the conversation, the General agreed that Tom's idea of slow disclosure could work, and that it was a great time for this to happen.

And, by the way, there's a lot of speculation about what he may have meant by that. This was right around the same time that US Navy pilots began to report hundreds of UFO sightings off the East Coast, so that could be it. Others argue that the proliferation of private companies in space makes it inevitable that someone will find something that makes disclosure a foregone conclusion, and that there are those in the government who would like to get out in front of that narrative.

Either way, The General agreed to help Tom find his advisors, connecting him with top-level officials and experts in the fields of space, intelligence, biological warfare, and even someone representing the President of the United states.

The C.I.A

During this time, Tom had also begun work on two book series: one fiction and one non-fiction. He claimed that both of these book series were heavily informed and influenced by his mysterious, high ranking advisors who had privileged knowledge of the UFO phenomenon, and who were giving him guidance and final approval as the books were being written.

Delonge managed to attach big names to these projects. The first of the fictional series entitled *Sekret Machines: Chasing Shadows* was written by *New York Times* best-selling author and Robinson Distinguished Professor of Shakespeare at UNC Charlotte, A.J. Hartley. And the foreword is written by Former Senior C.I.A Officer, Jim Semivan.

The first in the non-fictional series *Gods, Man & War* entitled *Gods*, was written by author Peter Lavenda, who is well-known for his books on occult history. And in a stunning coup, the foreword was written by the godfather of ufology himself, Jacques Valleé.

Shortly after the books were published in February and March of 2017, Delonge said that he was interrogated by the C.I.A. According to his accounts, the CIA caught wind of his books and that some of the things that he was describing, particularly with regard to secret government programs and technology, were so "on the nose" that it caught their attention.

Concerned that he truly did have high level advisors who could be leaking classified information to him, the C.I.A sent agents to San Diego to question Delonge. Tom claims that he was never officially detained, but that these weren't the type of people that you say no to when they ask to talk to you.

Over three days in a hotel, Delonge claims that he was interrogated by C.I.A agents about where he got the information from his books, and in the end determined that he was telling the truth—that his advisors had given him hints, but nothing classified, and that Tom had arrived at conclusions that so nearly mirrored reality by putting the pieces together himself.

Has Tom Delonge Lost His Mind?

So Tom is going around telling all of these stories about clandestine meetings with senior intelligence agents, high-ranking military brass, and the leaders of some of the most clandestine aerospace programs within the military industrial complex. And these claims about who his advisors are and how he found them, aren't even half as stunning and fantastical as what he says they've been telling him. (Which we will get to. I promise.)

As one would expect, the mainstream media mostly snickered if they paid attention at all. There were a few articles that came out in publications like RollingStone, but most seemed unwilling to print much more than the elevator pitch for To The Stars and preferred to dwell on the recent breakup of Blink-182 and the drama that ensued.

So most of these interviews were happening on far less mainstream outlets, including *Fade To Black* and *Coast 2 Coast*. While both of these shows are institutions within the UFO and paranormal communities, they don't have much credibility outside of those spaces—so these interviews were happening mostly off the radar.

However, even in front of what should have been a much more open-minded and sympathetic audience, the zealous and at times almost messianic tone of these interviews coupled with his extreme claims led most people to believe that Tom Delonge had lost his mind.

The Podesta Emails

But then something extraordinary happened.

During the 2016 presidential campaign, Wikileaks released over 20,000 emails from the DNC that had allegedly been obtained by Russian hackers. And among these emails was proof that Tom Delonge had in fact been in touch with high-level members of several government, military, and intelligence agencies and was being briefed on aspects of the phenomenon by these individuals.

The smoking gun can be found in emails between Delonge and John Podesta. Podesta has been an integral part of multiple presidential campaigns and administrations, having served as Chief of Staff to President Bill Clinton, Senior Councilor for the Obama Administration, as well as Hilary Clinton's campaign manager in 2016. Basically, this is a man who has spent more time in the White House than just about anyone, including Presidents.

It would be hard to point to a person with more power, influence, and connections in D.C. than Podesta. But Podesta was also known on the

Hill for having had a long-standing interest in UFOs. Upon leaving the Obama administration to work on the Clinton campaign he famously tweeted:

Finally, my biggest failure of 2014: Once again not securing the #disclosure of the UFO Files. #thetruthisstilloutthere @NYTimesDowd

When John Podesta's emails were leaked, it was revealed that he was one of the high-level officials who was regularly in contact with Delonge— confirming his claim that one of his advisors had a direct line to the President. And it was clear that Podesta had been briefed on Delonge's plan and was helping him to identify advisors and move his project forward.

I'd encourage everyone to take the time to read through these emails because they're honestly fascinating. And some of them can make you laugh out loud with the sheer absurdity of it all.

Here's one of my favorites:

On February 13, 2016, Delonge sent the following message to Podesta with the subject line "A good read…"

Here is the Digital copy of the Sekret Machines Novel.

I know you are so busy- So I apologize in advance. I ask that you consider reading my Foreword. I wrote this as a personal letter to the youth, so they can walk with me through all of this. When you read it, you will get that same sense- a wide-eyed and respectful experience. Also, you make an invisible appearance within the text at the end. :)

My Co-Author is a Distinguished Professor at the Robertson School of Shakespeare and NY Times best-selling Author….Basically, he's really f-ing good. Elevated writing, and award-worthy if he wasn't attached to my name, too. Ha-

You may actually love the book- If you like great stuff and have amazing taste. But until then, check out that foreword if you find time in all the madness.

To which Podesta replied a day later:

In Las Vegas for Hillary. Just did a taped interview with a local CBS affiliate. Got asked about the topic and gave an answer I think you would like. Hope they use it.

Wild, right?

But there are more shocking revelations in these emails than just a casual email bro-down between a man who has spent most of his adult career no more than a stone's throw from the leader of the free world and a rock star that most people who were still paying attention thought had totally gone off the deep end.

Because it turns out that basically all of what Tom Delonge had been saying about meeting with high-level officials was actually true.

Who Were Tom Delonge's Secret Advisors?

For example, Podesta's emails reveal that a meeting took place between Tom, Podesta and three other high-ranking individuals—and their identities give us some major clues to help fill in the blanks of his stories.

Aerospace Executive: Robert F. Weiss

The first was Robert F. Weiss, the Executive V.P. and General Manager of Aeronautic Advanced Development Programs for Lockheed Martin's Skunk Works. Sound familiar? There's no way to know for sure, but it seems likely that Weiss was the aerospace executive with whom he originally spoke and who introduced him to "The General".

And as for "The General" we actually have multiple suspects, both of whom were at this clandestine meeting.

Major General Michael J Carey

The first is Major General Michael J. Carey, the Special Assistant to the Commander of Air Force Space Command at Peterson Air Force Base in Colorado. Peterson Air Force Base is also home to NORAD (or the North American Aerospace Defense Command) as well as elements of the Space Force's Space and Missile Systems Center.

Major General Neil McCasland

McCasland is the former head of the Foreign Technology Back Engineering Laboratory at Wright Patterson Air Force Base in Ohio. McCasland also manages the Air Force's $2.2 billion Science & Technology program.

Wright Patterson Air Force Base has a long history of being at the center of speculation about the government's involvement with the UFO phenomenon. Most notoriously, Wright Patterson is where the army initially claimed to have sent the wreckage from the Roswell crash before changing their story and saying it was just a weather balloon. It was also the home of Project Blue Book.

Tom Delonge Is Vindicated

It seems likely that one of these two men is the General from Delonge's story, and regardless, the fact that both of them agreed to meet secretly with Delonge gives clear and undeniable credibility to his claims of having high-level advisors within the government.

And it's important to note that these weren't just any high-level advisors. In each of their official positions they are uniquely poised to have an insider's perspective on secret government programs involving UFOs. Basically, if there's anyone on the planet who might know the truth, it's the very people with whom Tom Delonge was meeting.

And whatever reasonable doubt remained was obliterated by what happened next.

The To The Stars Academy of Arts & Sciences

On October 11, 2017, Delonge announced the launch of his new venture, the To The Stars Academy of Arts & Sciences. The mission of TTSA expanded upon the earlier mission of To The Stars which sought to leverage media and entertainment to educate and prepare people for disclosure, to include ambitious research and innovation in the investigation of the phenomenon.

The new TTSA had three divisions: a science division devoted to further understanding of the phenomenon; an aerospace division devoted to developing new technologies and propulsion; and finally the entertainment division which would house all of Delonge's media projects.

Sitting behind Delonge at the launch press conference was the team that he had assembled and brought together to form TTSA, and the line up was, frankly, astonishing. Among them were some of the top names in intelligence, defense, and aerospace engineering, many of whom were going public for the first time.

So let's take a minute to meet some of the original members of the TTSA team who were on the stage that day:

Christopher Mellon | National Security Affairs Advisor

Chris Mellon served as Deputy Assistant Secretary of Defense for Intelligence for both the Clinton and Bush administrations. He also worked for many years on Capitol Hill as the Minority Staff Director for Senator John D. Rockefeller IV of the Senate Select Committee on Intelligence. As an aid to Senator William S.Cohen, he wrote the legislation that established the US Special Operations Command. He is a decorated member of the intelligence community, prolific author of articles on topics of national defense and security, and a political commentator. He received his MA in International Affairs from Yale in 1984.

Steve Justice | Aerospace Division Director

For 31 years, Steve Justice served as the Program Director for Advanced Systems at Lockheed Martin's Skunk Works where he worked on projects that in his own words "defy the imagination". He left this position to join TTSA and pursue disclosure, and to lead the development of emerging technologies associated with the phenomenon.

Dr. Hal Puthoff | Co-Founder and VP of Science & Technology

Dr. Hal Putthoff is a physicist and the Director of the Institute for Advanced Studies at Austin where he does cutting edge research on energy and propulsion systems. Since 1985 he has served as the President and CEO of Earthtech International Inc. And perhaps most interestingly, he ran the C.I.A. remote viewing program codenamed Project Stargate. This now declassified program ran for 20 years during the Cold War, as the C.I.A. allegedly used ESP as part of its intelligence gathering protocols. He is also a frequent advisor to NASA, the intelligence community, and corporations on leading edge technology and emerging technology trends. He holds a PhD in Electrical Engineering from Stanford University.

Jim Semivan | Co-Founder and VP of Operations

Jim Semivan is a retired CIA officer, working as a spy for over 25 years in the C.I.A. 's clandestine service and a former member for the Senior Intelligence Service. Since retiring, Semivan has served as a consultant to the intelligence community on classified topics including leadership training, tradecraft training, and programs for countering weapons of mass destruction.

Luis Elizondo | Director of Global Security & Special Programs

Yes, *that* Lue Elizondo—the former Director of the Pentagon's secret UFO program, AATIP, and who *only days before* the TTSA press conference had resigned from his position in protest of government secrecy regarding the UFO phenomenon. The exact timeline of events are unclear, but it seems likely that Delonge's project was a motivating factor in Elizondo making the move to abandon his career and pension to pursue disclosure.

Beyond his role at AATIP, Elizondo brings his expertise as a career intelligence officer who has worked with the Department of Defense, the Army, and the Director of National Intelligence. In his role as a secret agent, he conducted and supervised highly sensitive espionage and terrorism investigations. He also served as the Director of the Pentagon's Special Programs Management Staff, overseeing some of their most sensitive portfolios. He also has a background in microbiology, immunology, and parasitology.

So let's pause for a moment to take a breath and just take this in. Because this list of names is pretty staggering in terms of both their expertise and their proximity to privileged information about secret U.S. programs, particularly in the domains of aerospace and advanced technology—including the guy who ran the secret UFO program in the Pentagon for 10 years.

From that point forward, I think it's impossible to argue with any kind of credibility that Tom Delonge was lying about the high-ranking government and intelligence officials who had become his advisors. Not only did those advisors clearly exist, but they believed in Delonge and his mission enough to come out of the shadows—many of them putting themselves at great risk personally and professionally to do so.

Simply put—Tom Delonge was telling the truth.

Leaked UFO Videos & *The New York Times* Article

But Delonge still had one more ace up his sleeve, and he played it two months later.

On December 16th, 2017—the same day that the infamous *New York Times* article went live, exposing the secret government UFO program and sharing some of the leaked U.S. Navy footage of alleged UFOs—TTSA went live with the full footage of *Gimbal* and *FLIR*, the first two of the now famous videos.

It was these videos, along with the later released *Go Fast* footage, that began to wake the media up to the fact that this is a real issue, brought UFOs back into the public consciousness, and spurred legislation that forced the Pentagon to admit in 2020 that UFOs are, in fact, real.

The story of how the *New York Times* article came to be and who exactly leaked those US Navy videos is a little cloudy—most likely to protect the guilty, as leaking classified information to the press isn't exactly legal. But no matter the exact details or chain of events, it's hard to deny one thing—none of it would have happened without Tom Delonge.

Give Tom Delonge His Flowers, Dammit

And so that's where we'll leave it with Part 1. In Part 2, we'll get into where TTSA is in all of this now and we'll look at why it is that Tom Delonge has become such a controversial figure within the UFO community. And finally we'll take a—frankly, terrifying—tour through what Delonge claims his advisors have told him about the phenomenon, and how it's shaped his understanding of and beliefs about UFOs.

But for now, let's just take a moment to enjoy and appreciate the absurd beauty of this story, and how one man—through sheer grit and an abundance of passion—managed to turn the tide on the secrecy surrounding the phenomenon, forcing the Pentagon itself to admit a secret that it has guarded more closely than any other for the past 75 years:

That UFOs are real—and they're here.

7 MR. DELONGE GOES TO WASHINGTON [PART 2]

When we last saw our hero, Tom Delonge, he was basking in the glow of a seemingly impossible victory, not only proving that his shadowy advisors within the government were real, but leveraging those relationships to do something that no one has ever been able to do before—force the Pentagon to admit that UFOs are real.

Where Is TTSA Now?

So what's going on with Tom Delonge and his illustrious colleagues at the To The Stars Academy of Arts & Sciences now? Did they accomplish their mission?

That depends on who you ask.

On the one hand, since the formation of TTSA in 2017, the organization has succeeded more spectacularly in the disclosure effort than anyone ever has before. And in the history books, they alone will have the distinction of being the ones that forced the Pentagon to admit that the UFO phenomenon was real.

The importance and impact of that one accomplishment can't possibly be overstated. And the impact is still being felt.

However, I think that the public's reaction to these events—or more accurately, the public's lack of a reaction was as surprising to the members of TTSA as it was to anyone else who was actually paying attention.

There was the press conference! The *New York Times* article! The declassified videos! An actual admission from the Pentagon! And a report handed over to Congress that, admittedly didn't say much, but what it did say was Earth-shattering: that UFOs are both real and not human technology.

But somehow the Earth wasn't shattered.

These were barely b-stories in the 24-hour outrage cycle of the news media that, at the time, was deliriously glutted with the chaos of the pandemic and the election, breathlessly analyzing and pontificating on each and every one of the President's tweets. The revolution was here, the end was nigh, and all of it, apparently, was going to be televised.

A few weeks before I started this project, I started to ask my friends and family a question: "What would it take for you to take the UFO issue seriously?" I wasn't trying to change anyone's mind or to tell them what I believed. I just listened to what they had to say.

And of those I talked to, most of them said some version of the same thing which is summed up well by this comment from someone I know:

"I guess it depends on what you mean by UFO's. There are certainly things flying around that we can't immediately identify. If you are talking about the idea of beings from off the planet coming to visit, I guess as long as they choose to do it clandestinely, we will never know. Beings with sufficient technology to get here probably have sufficient technology to hide from us. If that is so, there is really no effect on my life or anyone else's. (If a tree falls in the woods and no one hears it, etc...) If they decide to reveal themselves it really won't be a question anymore, and the potential result is completely unpredictable. In the end there is really nothing actionable regarding the question, so it is hard for me to take it seriously."

I suspect that this is how most people think, and as a result, disclosure of the existence of UFOs—though undeniably monumental—had little to no impact on the public consciousness.

It's become clear that disclosure won't be a singular event, but a much longer and more complicated process that will unfold over time. Whether there ever is that earth shattering event that suddenly snaps the public consciousness into focus on this topic remains to be seen—but it certainly feels farther off than it did when TTSA took to the stage for the first time.

And TTSA itself has changed in both shape and substance in the years since.

In December 2020, Lou Elizondo, Chris Mellon, and Steve Justice left TTSA—though Jim Semivan and Hal Puthoff still remain part of the team. Still many perceived this exit as almost an admission of defeat for the organization—or at the very least, for Delonge himself.

For his part, Tom Delonge says that this changeup was partly due to the impact of COVID on his ability to attract investors to his, admittedly, high-risk projects. And in February 2021, papers were filed with the S.E.C. indicating that TTSA would be abandoning its science and technology initiatives and focusing solely on the entertainment division.

The split seemed to be amicable, at least, and there appears to be no bad blood among the original team members.

In an interview with George Knepp on *Coast To Coast* shortly after the split Lou Elizondo said:

"I love my friends at TTSA. They are incredible human beings, but I also have to say my mission has always been very clear, and that was to push disclosure forward. That's it. I think after three years, you know, I can look back and I think we've achieved much of what we've set out to do. TTSA, it's no secret, also focuses on its entertainment division and, let's face it, guys like Chris Mellon and Steve Justice and myself, we're not entertainers. So, very much like the History Channel project, we have accomplished our mission. Mission success. We have done more in three years collectively than anybody I think really expected us to achieve."

And I agree.

I think the only real fault you can assign to TTSA and what they've managed to accomplish is just that they didn't manage to accomplish all of the things that they set out to accomplish—but I think that bar is too high to be a fair one.

The original scope and substance of the TTSA mission was sweeping to the point of being grandiose. Among the many TTSA projects touted by Delonge was an effort to reverse engineer a UFO from recovered crash materials that he claimed to have in his possession.

And honestly, that just makes me smile. Because who would attempt such a thing? And of the people who would attempt it, who would say it aloud, knowing that they could still fail?

Tom Delonge—that's who.

And though his reach may have, thus far, exceeded his grasp, if it weren't for the absurd grandiosity of his vision, and his dogged, inexhaustible pursuit of it, none of this ever would've happened.

He may not be batting 1000, but anyone who would take a bet against Tom Delonge accomplishing whatever he puts his mind to simply hasn't been paying attention.

Why Is Tom Delonge A Controversial Figure In The UFO Community?

So, given all that we just discussed, you would think that Tom Delonge would be the undisputed hero of the UFO community. On the back of good old fashioned hard work and shoe leather, he somehow managed to penetrate the shadowy upper-echelons of the United States military and intelligence apparatus earning not just the support, but the apparent trust of some incredibly important people.

He assembled what can only be described as the *Avengers* of UFO disclosure who together managed to force Congress, the mainstream media, and the public-at-large to at least begin to recognize the reality of the UFO phenomenon and begin to reckon with the enormity of the challenges that lie ahead. And finally, as if that wasn't enough, his work directly led to the first ever declassified videos of UFOs being released and acknowledged to the public, along with the long-awaited admission that these craft are real.

And yet, the reality is that Tom Delonge remains a controversial figure within the UFO community. So what gives?

I think there are a few different reasons that this is the case.

He's Tom Delonge

The first is that some people are concerned about having Tom Delonge as the face of the UFO disclosure movement because they don't think that he always presents the community in the best light. And having spent decades fighting against stigma and ridicule, many in the community believe that we need to be putting forward leaders who are serious, credible, and credentialed in order to bring more legitimacy to the topic.

And the reality is that, despite being a wildly successful musician, singer, songwriter, author, record producer, actor, filmmaker, and serial entrepreneur, the image of Delonge and his bandmates streaking naked through the streets in the video for *What's My Age Again?* has been hard for Tom to shake—much to his chagrin.

And to be frank, despite Delonge's obvious desire to be taken seriously for his artistic, entrepreneurial, and even scholarly pursuits, he often hasn't done himself any favors in this department. He's seemingly never lost the conviction that he held since childhood that he could make his own way while being unapologetically himself.

Though clearly brilliant and staggeringly well-read, in interviews Delonge can sometimes come across as scattered, careening brazenly from one shocking statement to the next in a manner that can somehow come across as both overly casual and grandiose.

It can be hard to know what to make of him, especially when the things he says sound so entirely outrageous and over-the-top in their fantastic and conspiratorial nature that even among the UFO community there are certain segments that still believe him to be entirely unhinged. (And that's really saying something.) The fact that he's as likely to hit you with a dick joke as he is with the details of quantum mechanics or advanced propulsion systems doesn't help matters.

People Are Jealous

The second reason for Delonge's controversial status within the UFO community boils down to jealousy, or at least something adjacent to it. For those of us who believe that the study of the UFO phenomenon has the potential to unlock a new phase of human evolution—and I say "us" because I count myself among them—it can be hard to separate your desire to move disclosure forward from an all too human desire to be somehow *important* in doing so.

It's the very essence of what it means to be human to search for meaning and purpose in our lives, and what could be more meaningful than playing a role in humankind awakening to the reality that we are not alone—and perhaps never have been. It has all the makings, not just of future history, but of legend. It can be hard not to want to cast yourself in a leading role in a story like that.

And it's also important to recognize that there are so many members of the UFO community who have toiled in obscurity for years, if not decades—silenced, ridiculed, and ignored—who took on the Herculean and entirely thankless task of carrying the torch for this movement long before the Pentagon's recent admission on the reality of the UFO phenomenon made the subject at least somewhat palatable for the mainstream. And I think it can be tough for some people to have this cocky literal rock star—who looks and often acts like the skater dude in high school who would sell you drugs and then steal your girlfriend—come in and become not just the face of, but perhaps one of the most important figures ever in the disclosure movement.

Obviously this perspective fails to give Delonge the credit that he is due for imagining and then pulling off the impossible, but it's not hard to understand how people end up there or to have compassion for their stance, unfair as it may be.

He's Making Money

The third reason that some in the UFO community are suspicious of Delonge is that his work with UFOs is deeply tied to his business ventures. Delonge's vision for his role in the future of UFO disclosure involves a multimedia empire—books, albums, movie franchises, merchandise, and more. And for many, this alone is enough cause to question his motivations.

Within the UFO community there is a near ubiquitous suspicion of people who have monetized their work within the movement—and for good reason. As I'm sure you can imagine there are plenty of charlatans and huxters in the UFO world who exploit the unquestioning belief of some of the more vulnerable members of the community, claiming special knowledge and often the ability to facilitate contact between their followers and extraterrestrial or interdimensional beings.

Too often it starts innocently enough, selling books and training courses, maybe making a documentary or two, but somewhere along the way the whole thing takes a sharp left turn into cult territory. And suddenly they aren't just a researcher, but some kind of messiah, destined to usher in a new age of human evolution as we take our rightful seat as children of the cosmos or whatever.

It doesn't help that Delonge has at times aligned himself with some of these figures including Dr. Steven Greer—yes, the same Dr. Steven Greer with whom it is speculated that Delonge had his first encounter that caused him to quit Blink-182 to pursue disclosure full-time.

Although Greer was the driving force behind the notable but ultimately unsuccessful Disclosure Project of the early 2000's, he has earned criticism over the years for monetizing his CE5 techniques. For years, Greer has charged would-be experiencers for CE5 sessions in remote areas where they attempt to make contact with UFOs. He also offers training for people to become ambassadors who are then licensed to go teach CE5 on their own. Not only has Greer been persistently dogged by rumors that he fakes UFO encounters during his CE5 sessions (which happen almost exclusively at night) using flares, but his business model has often been compared to multi-level marketing—AKA a pyramid scheme.

Greer has not done himself any favors in the credibility department, making increasingly bizarre claims including that he was offered $2 billion by the government to stop talking about UFOs, which he allegedly refused. He's also come out strongly against TTSA, citing the counter-intelligence backgrounds of several of its members, including Lou Elizondo, and claiming that the work they're doing is a disinformation campaign that is likely a lead up to a staged alien invasion.

Anyway, I bring this all up as a side note, because I think all of this could explain why Delonge hasn't said that Dr. Greer is the ufologist with whom he had his first encounter with the phenomenon, even though it's fairly clear that it could be almost no one else. This could possibly be because of Greer's deteriorating reputation, and even more likely because of Greer's objection to his methods and the company he keeps.

But there's one more interesting possibility which is that, if it's true that Dr. Steven Greer does, in fact, fake UFO encounters for both clout and cash, wouldn't he be highly motivated to produce results for someone like Tom Delonge? You have to wonder if part of the reason that Delonge doesn't say his name is because it calls that whole story into question. Did he really have an otherworldly experience out there in the desert or was he the victim of a seasoned con man?

To be honest, it truly doesn't matter either way. None of that can diminish the importance and impact of the work that followed. Nor does that one individual experience have any bearing on whether or not the phenomenon itself is real. But it is interesting to think about.

Anyway, although there are plenty of people in the UFO community who aren't afraid to stretch—or even completely manufacture—the truth to make a buck, with an estimated net worth of $70 million, it seems unlikely that money is Tom Delonge's motivation. And when you're talking about funding everything from major motion picture franchises to reverse engineering exotic materials from UFO crash sites, it's possible—and even likely—that Delonge has sunk more money into this project than he's ever taken out of it.

Why Tom Delonge?

Another reason that some are suspicious of Tom Delonge is because they don't understand why, of all the people in the world, he was chosen to be the person to receive access to privileged information and allowed to be the face of what looks to be a sincere effort to pursue disclosure led by individuals within the government.

There's been tons of speculation about why it was him. Some people think that by making him the face of the project that they had plausible deniability if the whole thing blew up, and could easily throw him under the bus as just another delusional UFO kook. Others think that he's a "useful idiot" being used as the face of one of the largest scale disinformation campaigns in our nation's history.

But, at least in my opinion, no explanation seems to fit quite as well as the one given by Tom himself—that he was chosen because he was the one who showed up. Because he had some lucky breaks. Because he pushed the envelope when others would have shied away. Because he treated the topic and those involved with respect. But most importantly, because he came to them with a solution to a problem.

Everyone Hates The Military-Industrial Complex

And that right there is probably the biggest sticking point within the UFO community—that the pitch that seemingly got him in front of these high-ranking officials was one about helping the youth become less cynical about the military-industrial complex. We'll get more into the specifics of what Tom Delonge believes in just a sec—I swear—but suffice it to say that he sees the military-industrial complex as just one part of a necessary, and even heroic, international effort to respond to the biggest threat that our planet has ever faced.

As the General said, "It was the Cold War and they found a lifeform" and everything that they did after that, at least in this version of the story, was in order to protect the American people and allow them to live normal lives, while also moving heaven and earth to give them a fighting chance at fighting back against a potential enemy with technology far beyond our understanding.

And that right there, for many in the UFO community and elsewhere, is a cardinal sin—taking the side of the military industrial complex. I'll admit that, for me, it's a particularly hard pill to swallow.

But could there be something to what Delonge is saying—or at the very least, something we can learn from it?

I think that there is.

It's difficult to put yourself in the position of someone who at the end of WWII—the most horrible war our species has ever known—and in the early throes of the Cold War with the threat of nuclear war looming large in the public consciousness, had to make the decision about whether or not to tell the American people and the world that we'd not only discovered an intelligent lifeform, but one with capabilities beyond our understanding. I might even argue that it's impossible to truly put yourself in their shoes.

But after the events of the last few years, I think it's not too difficult to imagine what it was like to be just a private citizen during that time—the burning desire and the desperate need for stability after a time of crisis, for reassurance instead of despair, and the hope that life could one day be boring again. And we see that desire of the American people made manifest in the culture of the 50's.

And from that perspective, you can at least make a case for why the powers that be may have chosen to keep this quiet. And then when things happened that they couldn't explain—like, hypothetically, a UFO crash in the desert—they had to tell a lie to cover it up. And then another. And another.

Because that's the thing about lies—they tend to pile up. And when you break a rule once, it becomes easier and easier to justify doing it in the future. The road to hell, as they say, is paved with good intentions. Could it be that the culture of secrecy that grew up around the UFO phenomenon was put in place to protect us?

Perhaps.

But it's hard to deny that, even if the original intentions were good, at some point along the way things spun wildly out of control. Information about the phenomenon is jealously guarded by those at the highest levels of our government and stove-piped across different intelligence agencies, military branches, and even private corporations to an extreme that many question whether the President is even read into the most highly classified intelligence on UFOs, creating an illegal and undemocratic lack of oversight over something critical to our national security.

But it's more than just that. The government has lied to the American people, launched disinformation campaigns that ruined the lives and livelihoods of countless experiencers (many of whom are members of our own military), and, if we're being honest, probably much, much worse.

And on top of that, if the government has recovered exotic materials from UFO crashes, it seems likely that contracts would have been secretly awarded to select aerospace and defense companies, potentially giving them access to paradigm-breaking technology to the clear detriment of their competitors.

And many people's response to all of that is basically, "Fuck those guys. They lied. They broke the law. They broke their oath of office to serve and protect the Constitution. The UFO files need to be unsealed and those responsible for the cover up need to be held accountable." And, honestly, it's hard to find a flaw in that logic

The problem is that, faced with that sort of outcome, it seems unlikely that anyone who is in an actual position within the government to lead the charge on disclosure could be compelled to actually do so. They'd be effectively sending themselves and probably many of the people that they've known and worked with for years to the firing squad.

And beyond just what they'd be doing to themselves, there's what that kind of revelation has the potential to do to our country. One can only imagine the endless court hearings and tribunals as they try to decide who owns what and who is responsible and for what and how much? How long would that drag out and at what cost to our nation, both actual and spiritual?

Who would sign up for that? Who would make that call? And if you did somehow work up the guts to "do the right thing", would you know for sure that it was the right thing? I don't think it's that easy anymore. Too much time has passed. Too much has happened.

But what I see in Tom's story is not just hope—but a way forward. Because when offered just the tiniest sliver of hope that there might be a way out of the lies and secrecy without burning everything to the ground in the process, *look at how many people in powerful positions across the military-industrial complex answered that call.*

You don't need to believe what Tom Delonge believes or even to approve of his methods to recognize that his approach is, more than likely, a big part of the reason that he has been so successful.

What Does Tom Delonge Believe About The UFO Phenomenon?

And speaking of what Tom Delonge believes, it's finally time to talk about it. But to be honest, it kind of freaks me out. Like, the "makes me want to sleep with the light on" kind of freaks me out.

Award-winning investigative journalist and author of <u>In Plain Sight: An Investigation Into UFOs And Impossible Science</u>, Ross Coulthardt, perfectly explained why in an interview on *Theories of Everything* with Curt Jaimungal:

"I like to think of benevolent intelligence. I like to think that something that's vastly technologically superior to us and probably more intelligent than us has developed a moral value system that understands the importance of utilitarian ideas and moral decency and ethical values.

Tom Delonge freaked me out. I spend a lot of time listening to Tom Delonge interviews where he was talking about what he'd been told by the general and other people inside intelligence services before he went public with TTSA in 2016-2017. And, you know, I actually think it's really interesting because he talked about warring gods and jealous goods and I thought it was crazy.

And then the DNC leak happened. And it turned out that he really was talking—I mean it's beyond a doubt and people don't realize this, and again, it's just an amazing oversight by mainstream media that they haven't picked up on this. The leaked Wikileaks emails, leaked by the Russian GRU, believe it or not, who hacked the Democratic National Committee, who tried to gain intelligence that might help Trump win against Hillary Clinton—the DNC emails show irrefutably that Tom Delonge was telling the truth when he said that he was in communication with General Neil McCasland,

General Michael Kerry, Robert Weiss of Skunk Works Lockheed Martin, John Podesta, and other senior officers in Space Command, the Central Intelligence Agency, and different sections of the military were giving this punk rock star briefings, talking to him about what they knew.

What I find fascinating is journalists look for evidence, you look for corroborating evidence. And so I listened to Tom and I, like everybody else, I giggled. I thought, this guy's completely bananas. You know, there's one particular 'Coast To Coast' interview where I did a story on that. And then I remember feeling very humbled and quite ashamed of myself when I realized, oh my God, and I'm going through the DNC emails and going 'hells bells', if he was telling the truth about this, could it be he was telling the truth about the general when the general said 'it was the Cold War, and we found a lifeform'?

Could it be possibly true that the general did say that to him? And if he didn't, why would Tom lie about that?"

I remember listening to this interview the first time and feeling this weird but overwhelming sense of relief to hear someone as well-respected and clear-thinking as Ross Coulthardt say that Tom Delonge freaked him out. Because I felt the same way. I, too, was starting to wonder if I was crazy.

And it's not as simple as just "well he said these crazy things that turned out to be true, so all the crazy things he says must be true." A child can see through that logic. It would be extremely easy to dismiss any and everything that he said as delusional if that's all it was.

The real sticking point is that the people with whom he has been in contact were the exact people who would have real answers about the UFO phenomenon if there were any to be had. And these people not only briefed Tom on elements of the phenomenon, but several of them came out of the shadows for the first time—at great personal cost—to join him in his work with TTSA.

Some of these people, including retired C.I.A. officer, Jim Semivan, and physicist, Hal Puthoff, still work directly with Tom. And although some of the founding members of TTSA, including Chris Mellon and Lue Elizondo, have since moved on to pursue other disclosure efforts—and although they tend to shy away from some of Tom's bolder claims when asked—they have never denounced him and seem to remain on friendly terms.

And more importantly, they joined TTSA in 2017 at the time when Tom was giving his most unfiltered and shocking interviews. And yet they joined his organization and went on a media tour with him anyway, not departing the company until three years later. All of that seems highly unlikely if what Tom was saying about the nature of the phenomenon was simply the product of an unstable, paranoid mind.

So this leaves us with only two options, as far as I can see. Either:

1. At least some significant portion of what Tom Delonge believes is more or less true; or
2. This whole thing has been one enormous disinformation campaign meant to befuddle and confuse—to what end, we can't be sure

And listen, I know I've gotten very excited about this whole story, but that doesn't mean that I've entirely abandoned the possibility that this could all be a disinformation campaign—if only because given the two possibilities that I just mentioned, it's the only one that doesn't break my brain and make me want to never sleep again.

But I will say that if this has been a disinformation campaign, it's not been a very good one. Because, despite all the press and interviews and the legislation and the television shows and the admissions from the Pentagon itself, how many people are really talking about this?

Just doing the research for these two chapters took literal months to put together because it's scattered across the internet in podcast interviews, Youtube videos, books, rambling Reddit threads, and the occasional article from the "real media" that barely scratches the surface. Beyond a dedicated few in the UFO community, there's almost no one looking at the whole picture of this story and trying to figure what the actual fuck is going on here.

So if this was it—if this was their big plan for some sort of massive disinformation campaign, then where is it? They had a multi-millionaire rock star in the prime of his career with a growing media company at their full disposal and they did what with that exactly? If you asked ten people on the street who Lue Elizondo is, how many people do you think would know? Where I'm from, I'd say one—on a good day.

And what about all of these people in different parts of the government who have been working tirelessly for disclosure over the past 4 years since the New York Times article first dropped? Don't you think they would have gotten further in getting the attention of the greater public if they had actually had the full force of the United States government at their backs instead of in their faces actively trying to silence them?

Because, to be clear, there are those within the government who are working against them. There have been efforts to subvert and sabotage the Gillibrand Amendment, legislation that became law in December 2021 calling for the creation of an independent office to investigate UFOs, for the first time with the direct oversight of Congress and a board of mostly civilian advisors. And Lue Elizondo has been the subject of particularly vicious threats and attacks, and the Inspector General has opened a case to investigate the illegal and "mysterious" deletion of all of his records and emails from his time at AATIP in what can only be called a blatant attempt to discredit him.

In short, I find basically everything that Tom Delonge has said about the nature of the phenomenon to be very hard to believe. But I find it equally hard to believe, given the extraordinary and extraordinarily well-documented circumstances, that he's crazy or making it all up.

So—finally—what does Tom Delonge actually believe?

His views are as sweeping as they are bizarre, and I've done my best to collect them and put them into some sort of coherent narrative based on things he's said in interviews and what was written in his books. This list is probably not entirely accurate, as his views may have changed over time, and they are certainly not comprehensive.

But here we go. Here is a peek into the Universe according to Tom Delonge.

The Universe According Tom Delonge

Let's start at the beginning.

God is real—the summation of all the laws of physics and the source of consciousness. God is love.

And the Universe is teeming with intelligent life—some primitive and some more advanced than we can comprehend. Some flesh and blood, some inorganic, some artificial, and many with capabilities beyond our understanding.

There are several different groups and species of entities that are particularly interested in Earth and in humans. Are UFOs extraterrestrial? Ultraterrestrial? Interdimensional? Temporal? In a word—yes.

His advisors call them the Others.

One or more of these groups somehow meddled with the primates native to this planet to create humans. We're engineered in some way, perhaps even hybrids. But of what?

We don't know why there is so much interest in humans. Maybe there's something different about us. Maybe we were never supposed to be made. Maybe we are possessions and they are guarding us like livestock.

The Others have been here since the beginning of human civilization—which itself is much older than we had ever imagined. There have been many cycles of human history, each ending in calamity at the hand of the Others. What we consider to be "human history" is just the most recent cycle.

Atlantis was real—a great civilization that thrived because it had a way of defending itself against the Others. The Others eventually destroyed them for their hubris.

To me, the single most shocking thing that Tom Delonge has said is that in the beginning stages of meeting his advisors that it was his pitch that got him in the door, but it was something else that seemed to be the key to these people taking him seriously—

And that was the idea that the entire UFO phenomenon is about multiple "gods with a little *g*" that fight amongst themselves and by design factionalize mankind into different religions so that we fight each other—to what end, we don't know.

It might be some kind of proxy war, it might be for entertainment, or—as Delonge has often suggested—they might actually feed off the energy of human suffering.

During WWII, Melanesian islanders were exposed to modern technology for the first time. And, despite the raging war, the vast amount of food, equipment, and medical supplies that were airdropped there significantly improved the lives of many of the islanders. As a result, some of the islanders began to worship the planes that would do these airdrops, thinking that they were sent from the gods.

This phenomenon is called a "cargo cult" and according to Delonge, that's essentially what all of the major world religions are—cargo cults.

Virtually all of the ancient religions and traditions talk of the time when gods walked among us or at least regularly made their presence felt—

and that actually happened, but it wasn't gods walking among us, it was the Others.

Our government is aware of this.

The Roswell crash was the first time that we got our hands on this advanced technology. This changed everything.

Within three months of the Roswell crash, the Air Force was separated from the Army, the Department of Defense was created, and the C.I.A. was created, to investigate and potentially defend the country against this new threat. But this also allowed them to stovepipe information so that only a few people at the top had the full story.

At nearly the same time in both Russia and China there were similar crashes. Tom says this was intentional. The Others seemingly gave this technology to the three great world powers at the same time. Why? Tom has speculated that it could be to encourage more war, or it might just be to see which of us is strongest.

The government hid what they knew from the public because they were also trying to hide what they knew from the Others. They built bases and testing facilities deep underground, and grimly, heroically, got to work

The Cold War served as the perfect cover for decades of exorbitant spending and the stockpiling of advanced weaponry. The United States and the Soviet Union were engaged in an intense standoff, but the reason that the Cold War never got hot was because behind the scenes they were working together against the Others.

To protect this secret they launched a massive disinformation campaign, starting with Project Blue Book. With Project Mockingbird they infiltrated the media to turn the narrative to make the subject of UFOs taboo—something that only kooks and conspiracy nuts would ever take seriously.

They started the conspiracy theory that the moon landing was fake to distract from the truth—that the Apollo astronauts encountered UFOs and intelligently built structures on the moon.

The Others have bases on the Moon, in our oceans, and even underground.

Abductions and cattle mutilations are real, and these phenomena are closely tied to stories of demons. These entities are one in the same.

There is at least one group of the Others that hates humans—that would destroy us if they could.

This may be because we have a soul—and therefore, a direct link to the loving, eternal source that is God. Perhaps these beings don't have that. Perhaps advanced forms of AI don't have a soul. Tom has suggested that the Grays may be artificial drones, and that transistors have been seen on the backs of their heads.

Or maybe it's more than just jealousy.

There is something about the UFO phenomenon that is deeply tied to consciousness. Perhaps humans are much more powerful than we ever imagined. Perhaps we've forgotten who we are. And maybe there are beings who will do anything to prevent us from remembering.

Delonge says an advisor told him that there's a black pyramid underground in Alaska. They discovered it when anomalous measurements showed up during an earthquake. It's twice the size of the Pyramid of Giza and they've been studying it for years.

"We've finally figured out what it is," Delonge's advisor told him. "It's suppressing human consciousness."

So, What Now?

I could go on, but I think you get the idea. It's *a lot*.

Which is why it's, frankly, kind of distressing that so much of what Tom Delonge has said has checked out and that so many of his advisors, not only have been proven beyond a shadow of a doubt to be his advisors, but are the very people within our government who would know whatever secrets they may be hiding about the UFO phenomenon.

I'm not saying that everything that Tom Delonge is saying must be true—but at least some of it almost has to be. And any one of the things that I just told you would change everything. Taken as a whole, it makes me wonder if anything that I once believed about the nature of my reality is true.

And that's the real takeaway for me here—that when it comes to the true nature of the UFO phenomenon, literally everything is on the table. And it's clear that, whatever the phenomenon is, the truth is sure to be stranger than fiction, and will undoubtedly challenge our ideas about who we are and what place we hold in the cosmos.

So that's where we'll pick up next time, by beginning to follow some of these threads of possibility to see where they might lead us. And we'll begin this journey at the dawn of human civilization itself.

8 A Rational Approach To Ancient Aliens [Part 1]: Archeology & Epistemology

In this chapter, we begin a new leg of our journey down the rabbit hole, so I want to take a quick minute to recap where we've been and to begin to map out the way forward from here.

So far we've established that the Pentagon has admitted that UFOs are real, that they are operating with impunity in our airspace, and that we don't know what they are. We've talked through some of the most common explanations for what the UFO phenomenon might be, considering everything from secret human technology to future humans and from extraterrestrials to ultraterrestrials.

We've also been introduced to some—though certainly not all—of the main players in the current disclosure movement including former Director of AATIP, Luis Elizondo; former Deputy Assistant Secretary of Defense for Intelligence, Chris Mellon; investigative journalist, Ross Coulthart; former Blink-182 frontman and founder of the, now defunct To The Stars Academy of Arts & Sciences, Tom Delonge—and many more.

Although we've covered a lot of ground, when it comes to understanding the full scope of the UFO phenomenon and its implications for humankind, we've barely scratched the surface. And yet, already it's clear that the branches of ufology stretch out in every direction. Its tendrils are deeply tangled and entwined with questions of human history, consciousness, spirituality, quantum mechanics, and even the paranormal. Basically anywhere that we would look to ask the deeper questions about the nature of our reality and our purpose within it, we find ufology lurking—taunting us with seemingly unanswerable questions.

And so moving forward in this book series, we'll begin to get a high-level overview of the many different facets of this phenomenon. We'll talk about bleeding-edge scientific discoveries, dive deeper into the history of UFOs and what the United States government might be hiding, explore the deep ties of the phenomenon to the paranormal, discuss multiple models of the mind and consciousness, and even look at the mysteries surrounding the origin of the human species and what makes us so unique.

UFOs & The Dawn Of Civilization

For now though, I want to talk specifically about some of the mysteries that exist around the dawn of human civilization—where it happened, when it happened, and why it happened—and what role the UFO phenomenon may have played in shaping who we are today.

As we've discussed previously, it's basically impossible to point to any specific incident or time period and say that that is the definitive beginning of the UFO phenomenon. The reality is that, although it has manifested itself in different ways throughout history, there is a clear pattern of strange objects and beings coming from the sky being reported by people and cultures around the world, dating back to our very earliest written records.

There are lots of explanations for what these reports might represent—from misunderstood weather phenomena to myths and allegories that were never meant to be taken as fact. And because it's very difficult for us to know exactly what happened hundreds, if not thousands, of years ago, it can be very tempting to just write all of those stories off as figments from a more primitive, less scientific time period.

Evidence Of UFOs & Aliens In The Past

However, with the Pentagon's admission that UFOs are real, I think there is a strong argument to be made that we no longer have that luxury. Members of our military and witnesses all around the world have reported seeing strange things in the sky that don't abide by the laws of physics as we know them—

why would we assume that this is the only time in human history that this has happened?

In fact, when we examine the hypotheses offered to explain the UFO phenomenon, most of them function in such a way as to strongly imply that this isn't a new phenomenon.

For example, if what we're dealing with is Future Humans, then there's no reason to believe that they'd only come back to this time period. If they're going back to observe or interact with earlier versions of humans at all, we can assume that this has been happening throughout human history.

If what we're dealing with is Ultraterrestrial, then whatever this is has been here for a long time—potentially much longer than humans.

If what we're dealing with is Interdimensional—well then there is almost no limit to who or what is coming through the veil, and limiting humanity's exposure to these beings to just a few decades is nonsensical. If interdimensional travel is possible at all, it's almost certainly happening on a scale that we can scarcely comprehend.

So of the most common hypotheses, that really only leaves the extraterrestrial hypothesis. Could beings from another planet have first arrived here about 80ish years ago and have been messing with us ever since? Potentially. If we're limiting the scope of the phenomenon to the last century, then it's certainly the hypothesis that makes the most sense.

However, if extraterrestrial visitation is happening now, it doesn't preclude extraterrestrial visitation in the past. If Earth and humans are interesting to alien intelligences now, it's likely that they would have been in the past, as well. Though, admittedly, the advent of the nuclear bomb could conceivably have been the thing that put us on their radar. It's hard to know.

The point being, given what we know of the phenomenon so far, it makes a lot of sense to look to the past for potential clues.

And you know what that means, friends—it's time for us to do an *Ancient Aliens*.

What Is The Ancient Astronaut Theory

For those who aren't familiar, *Ancient Aliens* is a show on the History Channel that is based on the premise that in our distant past, extraterrestrials came to Earth and interacted with—and sometimes helped—humans.

This theory is called Ancient Astronaut Theory and its proponents see evidence to support their claims in everything from petroglyphs depicting strange beings to ancient megalithic structures that they contend humans could not have built on their own. They see the stories in ancient scriptures as evidence of our ancestors trying to describe events and technologies that they had no words or framework to understand. Basically anything interesting, mysterious, or anomalous in our past has one explanation—aliens did it.

Or at least that's the main takeaway of the show. *Ancient Aliens*, by it's very infotainment nature, is not big on nuance and is notorious for stringing together wild theories and making bold proclamations based on evidence that is dubious at best. It makes for a fun watch if you're not too concerned with the truthiness of the facts being presented. But for those who are looking for a scientific, rational approach to some of these bigger questions—this isn't it.

In Defense Of Ancient Aliens

Now I want to come right out and say that I am an *Ancient Aliens* apologist. There are many in the UFO community who don't share my views—and I understand why.

For many people, *Ancient Aliens* is their main—and perhaps their only—exposure to ideas surrounding the UFO phenomenon. I was certainly one of those people before I started down this rabbit hole. And if that's the main place you're getting your information, it can be hard to take any of it seriously. It just comes across as totally wackadoo.

But I still go to bat for *Ancient Aliens*.

In my early days of investigating this phenomenon the deal that I made with myself was that I would look at every piece of evidence and every theory. I would turn over every rock, and I would allow myself to really and truly consider ideas that I'd deemed to be absurd in the past. I didn't need to accept it. I didn't need to believe it. I just needed to explore it with an open mind, doing my best to set aside my previous biases and weigh each piece of evidence on its merits.

And being new to the world of UFOs and not knowing where to start, *Ancient Aliens* was a pretty natural place to begin. And so I started watching. And in the midst of the logical leaps and the hilarious non-sequiturs, there were also some truly astounding and challenging ideas. There were revelations that, frankly, floored me—prompting me to dig deeper.

The ideas that really struck me—the ones that kept me awake puzzling over if this could possibly be true, and contemplating the stunning implications if it were—weren't the flashy ideas. It wasn't wild speculation about nuclear wars or alien overlords or high technology in the distant past.

It was the smaller and more grounded details—the physical proof, the anomalies that could be ignored, but not denied, the tantalizing hints that the story of humanity might not be what we think it is.

And I'm grateful to *Ancient Aliens* for that.

Other Explanations

However, it's also important to recognize that although the ancient astronaut theory is the idea that most people are familiar with, there are actually a few other potential explanations for the mysterious sites and anomalous artifacts from our distant past—and most of them boil down to the idea that human civilization may be much older than we thought.

We'll get back to that idea in the next chapter and talk through some of the evidence for that potentially being the case. But for now, let's stick a pin in it. All that's important at this stage is to recognize that there are a wide range of possibilities when we look at the mysteries of our past—and many of them are just as startling and just as profound as ancient alien visitors.

The Big Question: What Do We Know For Sure About Ancient Human Civilization?

But before we start looking at any particular evidence or explanation, we first need to ask the biggest and most important question—what do we know for sure about ancient human civilization?

After all, whether we're talking about alien intelligences interacting with and influencing human development or human civilization being older than we think it is—there should be evidence, right? And with the technological advances of the past several decades, archeologists and researchers are equipped with more powerful and precise tools than ever before. So surely there must be some things that we know for sure, which means that, hypothetically, there are things that we should be able to rule out, right?

The answer to that question was meant to be just the intro of this chapter, but has ballooned into a chapter of its own—and I think it's important that we spend some time here. Because before we can have an intelligent and informed conversation about what may or may not have happened in humanity's ancient past, we first need to have the conversation about what we already know and how we know it.

And also, as you've probably already guessed, I'm going to advocate for the idea that there is significant evidence that the established narrative about how, when, and why human civilization developed is incorrect—which is not something that I do lightly. I am not an archeologist or a scientist. I'm a content creator who reads too many books. And no matter how many books I read, any decent archeologist has forgotten more about this subject than I am likely to ever know.

So I think it's important to first lay out the case for why I believe this line of questioning is warranted, and why I feel comfortable making the admittedly audacious claim that the experts may be wrong on this one.

Intro To Epistemology

Before we get into all that, I want to take a minute to talk about a very important concept that we'll inevitably be revisiting again and again as we continue our journey down the rabbit hole—and that concept is epistemology.

So, for those of you that didn't waste your parents' money on ¾ of a philosophy degree, epistemology is the study or theory of the origin, nature, methods, and limits of knowledge. As we continue to explore all these strange ideas and theories, we'll inevitably find ourselves coming back to the same basic questions:

What do we know to be true? And how do we know for sure that what we accept as true is actually true? Basically, it's exactly what we're talking about with regards to early human civilization, *what do we know and how do we know it?*

These may sound like really obvious questions with equally obvious answers, but the question of what we can know and how we can know that we know it has stumped philosophers—perhaps more than any other—for thousands of years.

Consider for a moment the difference between *believing* that something is true and *knowing* that something is true. We know instinctively that those two things are very different, but it can be hard to put your finger on *how* they are different.

A person can believe that something is true that is actually false. This happens all the time, often when someone is working with incomplete or incorrect information. People can jump to conclusions or put blind trust in a source that isn't accurate. Sometimes it is just a matter of a mistake or a miscommunication. But generally we agree that just because someone believes something, doesn't necessarily mean that it's true. And we also recognize that we can become attached to our beliefs in a way that makes us more likely to dismiss or overlook data that doesn't fit within that framework.

But knowing something is fundamentally different from believing it. You can know your best friend's phone number. You can know that 2+2=4. You can know a poem word-for-word. You can know your name. And inherent in this idea of "knowing" something is the assumption that this knowledge is both objectively true and verifiable.

Simple, right? Not quite.

We don't know anything in a vacuum. Every piece of knowledge that we have, is built upon a foundation of other data and assumptions that support and confirm it—even if you never think about it. You don't just know that 2+2=4. You know what "2" of something is. You can picture it in your mind. You know what a whole number is. You know what the "+" and "=" mean, and you can perform those mathematical functions in your head.

But the number 2, like all numbers, is itself an unprovable abstraction. Using number theory and this shared understanding that we have of what numbers are, we've created logic—which itself is the fundamental underpinning of science and language, and basically all of the cool stuff that humans know how to do. But while we can use numbers to do all of that, we can't use numbers to prove themselves.

You could say that 2=2, but saying that two equals itself is not a meaningful statement in this context—just like if you said "blue is blue".

Neither is saying that 3-1=2. We can prove that that equation is consistent with the framework of mathematics, but we can't prove that it is real *outside* of the framework of mathematics.

Now, I don't want to get too bogged down here, because this line of questioning inevitably leads to the question of whether we can actually know anything at all. If you drill down far enough into any single thing that we "know" to be true, you inevitably hit conceptual bedrock—or the place where your verifiable knowledge ends and the unverifiable, abstract concepts and assumptions from which they grow begin.

Godel's Incompleteness Theorems

Logician, mathematician, and philosopher Kurt Gödel first expressed this idea with his Incompleteness Theorems in 1931. But you don't have to be a mathematician or a philosopher to understand the limitations of our knowledge. If you've ever spent any time around a young child, you likely already know that each of us is only about 5 "whys" away from a complete existential unraveling.

Why is grass green?

Because it has a bright pigment called chlorophyll that makes it green.

Why?

Because plants need it to make food in a process called photosynthesis.

Why?

Because that's how it evolved over billions of years.

Why?

God maybe? Or maybe it's just the random unfolding of events over billions of years. I honestly don't know, kid, and I don't have time to be contemplating the absurdity of existence right now. Put your shoes on.

Now, for our purposes, it doesn't serve us to get lost in navel-gazing questions about whether or not knowledge is even possible. We conduct our lives under the assumption that we can know things, and our observable reality seems to suggest that we can. After all, our knowledge allows us to do things that we otherwise wouldn't be able to do, like calculate the exact angle of entry for a spaceship returning to Earth so that it doesn't burn up in the atmosphere.

So if our experience suggests that knowledge is possible, why am I bringing this up?

I bring it up because I think that it's important to bring a certain level of humility to any conversation about what we can know and what we can't know. We have very good and very reliable frameworks that allow us to understand complex ideas and relationships, and methodologies that allow us to verify our conclusions against the conclusions of others. But it's important to remember that these things are tools, and not necessarily a representation of an objective, universal reality.

This is a mistake we make a lot, as humans.

Science Isn't "True"

Neil deGrasse Tyson has famously said that, "The good thing about science is that it's true, whether or not you believe in it." And given the cultural context of the last few years, many champions of science have grabbed onto this statement and have used it as both a rallying cry and as a weapon against those who don't share their views.

But if you break it down, that statement isn't just blatantly false, but fundamentally unscientific. Science isn't "true". Science isn't a collection of verified facts. Science is a methodology, and like any methodology it's as fallible as the people who are utilizing it.

Returning to an example that we've used a few times now, let's consider 2nd century mathematician and astronomer, Ptolemy. His model of the solar system improved upon past models so significantly that it was used to accurately predict the movements of celestial bodies in the sky for over 1400 years—but it was also completely wrong and based on the assumption that the Earth was the center of everything.

To be clear, I'm not saying that we shouldn't trust scientific conclusions. The scientific method is *by far* the best way to test our assumptions about our reality in a way that is as free from bias as possible. It's imperfect, yes, but we used it to put a man on the moon. So it's still pretty damn amazing.

But what I am saying is that we need to be careful not to turn science into a religion. There are no sacred cows in science. There is no scientific "fact" that can't be undone with the introduction of a new data point that changes everything.

But that's a hard thing for humans to remember. We live our lives with the assumption that we can observe and have knowledge of objective reality—which just makes practical sense. It keeps us safe. It creates order where there was none. It's the basis of civilization itself.

So it can make us super uncomfortable to recognize that our sense of objective reality is so contingent and potentially tenuous. We don't like to think about the fact that everything that we regard to be true is essentially just a hypothesis that is waiting to be disproven.

We don't like "I don't know." We like answers. We like to believe that those answers are clear. We like to believe that our species has cracked everything from the structure of atoms to the limits of the Universe and that there is nothing left for us to know. And when we have questions about our reality, we increasingly look to scientists to be the arbiters of truth.

But scientists don't have answers. They have science—a methodology that allows us to approach the truth, but is not truth itself. And it's not just that their answers *can* change—given enough time, they *will* change.

A pattern that we see repeated again and again throughout history— almost without exception—is that when we get new data that upends our models and invalidates our fundamental understanding of the nature of our reality, that data is initially rejected as impossible, ridiculed as pseudoscience, and marginalized along with all those who dare to entertain such heretical viewpoints.

And seemingly no one is totally immune from this bias against data that threatens our worldview. Einstein himself said that the greatest blunder of his career was not believing what the math was telling him about the nature of the Universe—that it wasn't static and unchanging, but was actually expanding at an accelerating pace. Einstein invented the cosmological constant to normalize his equations—and it wasn't until Edwin Hubble proved that the light from distant galaxies was red-shifted away from the Earth that he realized he'd been wrong all along.

And I would argue that—as profound as Einstein's discoveries about the Universe were—the idea that we are not alone, and perhaps never have been, represents the single greatest paradigm shift in human history. The implications are as challenging as they are wide-reaching, and we should expect that it will make people uncomfortable. It will make people angry. It will make people defensive.

Because, to be clear, what we're talking about is nothing less than rewriting human history as we know it. But sometimes that's exactly what science requires of us.

Archeological Epistemology

Alright, I think we've spent enough time in the philosophical weeds. Let's get to talking about the matter at hand, which is—*what do we know for sure about early human civilization?*

As we'll see this isn't the easiest question to answer, at least in part because archeology itself presents some unique epistemological challenges and complications.

How Do Archeologists Know How Old Something Is?

Let's start with dating. If you want to understand what happened in the distant past, you need to be able to construct some kind of a narrative. And to construct an accurate narrative, you need to know not just when things happened, but the order in which they happened relative to other things.

So how do researchers figure out how old something is?

There are several different methods depending on what it is that you're trying to date, and they fall into two main categories: relative and absolute.

Relative Dating Techniques

Relative dating techniques involve establishing a basic timeframe for something by comparing it to other old things—and are therefore far less precise. It's actually more accurate to think of these methods as "ordering" rather than "dating".

Before more precise methods of dating were available, relative dating techniques were the main way that researchers attempted to assign dates to things—and although they aren't as precise, many of these techniques are still used today.

Biostratigraphy

One such method of relative dating is biostratigraphy. As rock and sediment build up over time, they form layers. This means that, in most cases, you can assume that if you find a fossil or an artifact in a certain layer, it's older than anything found above it.

Within biostratigraphy is a submethod called faunal association. Sometimes researchers can compare things to other fossils that were found in the same layer. If you know how old those fossils are, then you can establish that what you are dating was around at the same time.

You can potentially get more precise dating information by looking specifically at microscopic animal life in the fossil record. Microfauna tend to evolve much faster than larger organisms, so each species exists for a much shorter time in the fossil record, allowing researchers to zero in more precisely on a particular time frame.

Paleomagnetism

Another method of relative dating is paleomagnetism. About every 100,000 to 600,000 years, the Earth's magnetic poles flip. These changes can be detected by looking at the orientation of magnetic crystals in certain kinds of rock. Now obviously, 100,000 to 600,000 is a pretty long timeframe making this method not very precise, however, it is often used as a check against other methods of dating to help confirm their conclusions.

Tephrochronology

Researchers can also use a method called tephrochronology. Whenever there is a major volcanic eruption large amounts of dust, rock, and other materials are shot up into the atmosphere where they eventually rain down on the land below. This layer of sediment will have a unique geomagnetic fingerprint. So if you know the date of a particular eruption, you can date things relative to that layer with everything above it occurring after the eruption and everything below it occurring before.

Absolute Dating Techniques

As you can see from these methods, they aren't particularly precise. However, scientists have also developed several other absolute dating methods that allow them to get a much more accurate sense of the age of something.

Radiocarbon Dating

The first, and by far the most common is radiocarbon dating, which involves measuring the quantities of carbon-14. Carbon-14 forms high up in the atmosphere and is then breathed in by plants and breathed out by animals. As a result, you can find carbon-14 in anything that is alive.

Carbon-14 is a radioactive isotope. A radioactive isotope is basically a version of an atom that has a different number of neutrons. Carbon usually has 6 neutrons, but carbon-14 has 8, which makes it both heavier and less stable.

Because it's less stable, carbon-14 breaks down over time, with one of its neutrons splitting into a proton and an electron. The electron then escapes, while the proton stays, and with one fewer electron and one more proton, carbon-14 decays into nitrogen.

This process begins as soon as something dies and stops taking in carbon-14, but the process of radioactive decay is very slow. Its half-life—or the amount of time it takes for half of a given quantity of carbon-14 to decay—is 5,730 years. Scientists are then able to use the amount of carbon-14 left in organic materials, like bones or plant fibers or ashes from a campfire, to figure out how old something is.

However, there are limitations to radiocarbon dating. The first is that after a period of about 50,000 years, organic materials have lost more than 99% of their carbon-14, which means that we can only use it to date things from the last 40,000 years or so. The other limitation is that it only works on organic matter, so you can't use it on things like rock, metal, or other minerals.

Single Crystal Fusion & Uranium Series Dating

To date rock, researchers usually turn to either single crystal fusion or uranium series dating. Without getting too in the weeds, like radiocarbon dating, both of these dating techniques involve using the decay rates of various isotopes to determine how old something is. Conveniently, while uranium series dating only works for things that are 40,000 to 500,000 years old, single crystal fusion works best on things that are 500,000 years old or older—and actually gets more accurate over time.

Trapped Charge Dating

For materials like teeth and coral that are especially good at trapping electrons from the sun and cosmic rays, researchers can use a technique called trapped charge dating. This can be a complicated process that involves looking at multiple variables, including the amount of radiation that the object was exposed to each year, to calculate the rate at which electrons were trapped. And to make things more complicated, it's only accurate for things that are less than 100,000 years old.

Thermoluminescence

Besides teeth and coral, certain silicate rocks, like quartz, are also very good at trapping electrons. For researchers who are specifically working with prehistoric tools made of flint—which is a hardened form of quartz— thermoluminescence can help them determine the age of the tool itself—not just the materials it was made from. The last stage in the process of making these tools was usually to drop them into a fire, which frees all of the electrons, essentially resetting the clock for that object. This allows researchers to determine how long ago it was made.

Optically Stimulated Luminescence

For things that have been buried for a long time, researchers can use Optically Stimulated Luminescence in a process that allows them to determine, not when something was made, but how long it's been since it was last exposed to sunlight.

Good Dates, Bad Dates & Ugly Dates

There are a few other methods of dating, but I think you get the idea. You don't need to know every method of dating or how it works. What's more important is to understand the challenges that researchers encounter when they are trying to date an archeological find.

Each method has its strengths and its limitations. And because of these limitations, researchers generally strive to date things in as many different ways as possible to help them zero in on the correct date. The result is that there is a lot of variability in how accurate dating of these objects can be.

If something is dated using at least two different methods and is verified by multiple independent labs who all point to the timeframe, you can have a pretty high degree of confidence that that date is legit. If something is only dated using one method or if it isn't independently verified, that date may be less reliable.

Challenges Of Dating Megalithic Structures

And then there is the case of the things that you can't date accurately with any of the above methods—and maddeningly, this applies to a category of archeological treasures that we'd most like to date—which is basically any structure made out of stone.

Thousands of years ago, cultures around the world erected massive stone monuments, temples, and other structures out of enormous blocks of stones referred to as megaliths. Many of these megalithic structures remain largely intact today, even as the civilizations and cultures that built them have been lost to the sands of time.

Their stunning architecture stirs our sense of wonder. Their confounding size pricks at the corners of our imagination. Their mysteries call to us, and there is this uncanny sense that if only we could unravel their secrets, we might gain some special insight or knowledge that somehow has been lost.

They're also extremely difficult to date.

We can tell when the stone was formed, geologically. We can make some assumptions about what kinds of tools would have been needed to build these structures that can help us pin it to a particular time period. We can compare it stylistically to other art, architecture, and writing that we *are* able to date and draw some conclusions that way. And we can often date organic and other materials found in and around the site.

But all of these are indirect and imprecise ways of arriving at a date. And if you're working with the wrong set of assumptions to begin with, you could misdate something entirely. It's a messy business.

For instance, you could find a crypt with bones in it and date the bones to a certain time period, and then assume that the construction of that crypt was more or less contemporaneous to the people who were buried there. But what if it had been in use for thousands of years before that? What if it had changed hands a few times due to war or even just the passage of time?

The most educated guess in the world is still just a guess without anything to confirm it.

Many megalithic structures are built on top of earlier structures.

Another challenge in dating megalithic structures—or really any structure built in antiquity—is that cultures around the world have shown a tendency to rebuild on top of an existing structure or foundation. You see this especially in places that have a religious or spiritual significance.

There's perhaps no place on Earth where we see this occurring more than in Jerusalem, where three different monotheistic religions—Judaism, Christianity, and Islam—have some of their most sacred sites. And in some cases, these sites are literally right on top of each other.

The Islamic holy site The Dome of the Rock sits on Temple Mount in Jerusalem over a spot that is believed to be both the place where the God of the Old Testament asked Abraham to sacrifice his son, and the spot from which Muhammad ascended into heaven. The Dome of the Rock is built on the former site of the Second Jewish Temple, which was first built in 516 B.C., which in turn was said to have been built on the foundation of Solomon's Temple which was said to be completed around 957 B.C. The Western Wall is believed to be a remnant of the second temple.

So in this one spot you see layers upon layers upon layers of history. And this creates complications. Obviously, we can't really go digging around these sites to try to learn more about what's under them. Beyond their profound significance to billions of people around the world, they are archeological treasures in their own right. Doing anything that could potentially damage these structures is unthinkable.

And because we aren't able to study what lies underneath these sites, the evidence for any older structures that may have come before is lost to us—their mysteries and meaning swept away by the sands of time.

Megalithic structures can last for thousands, if not tens of thousands, of years.

Another thing that makes megalithic structures particularly difficult to date is that they are built to last for a really long time—far longer than anything else built by humans.

To help put that into perspective, you can think about it this way: If every single last human alive disappeared from the planet tomorrow, within just 1,000 years almost all traces of our civilization will have been eroded, buried, or otherwise swallowed by nature. Even Manhattan will once again be a lush, green island—much like it was when it was first explored by Henry Hudson in 1609.

Some signs of humanity will still remain though. The Washington Monument will likely still be standing, though it would likely be underwater. The walls of Notre Dame in Paris may still be recognizable. Stonehenge will still stand

But if you zoom out to 10,000 years after humans, what would remain then?

The Great Wall of China will have eroded tremendously, though it will still be recognizable. The pyramids and the Sphinx will likely be long gone, unless they are buried by the Sahara. Only our largest stone structures would remain—like the remnants of the Hoover Dam. The faces on Mount Rushmore, however, could survive in a recognizable form for millennia.

This is because granite, one of the hardest kinds of rock and a popular megalithic building material, erodes at only about one inch every 10,000 years. So when we're talking about large structures, built with granite (or similarly durable) blocks of stone, we're talking about something that has the ability to exist and endure outside of the normal human time scale. In the time it would take for New York City to become a forest again, a megalith will have undergone a bit of weathering.

When we're talking about structures that have the ability to endure for thousands of years, if not tens of thousands of years, and without any direct way to confirm when they were built, it's entirely possible that our understanding of how old they are is entirely wrong. And if, as some archeologists suggest, some of these megalithic structures are older than we previously thought—what might that mean for the story of humanity? And what might we discover about ourselves as we begin to reimagine the events of our distant past?

Other Epistemological Challenges Of Archeology

And, beyond challenges with dating the most interesting and mysterious remnants of our ancient past, archeology itself presents some additional epistemological challenges as a result of the unique way that data is collected in this field of science.

Archeology often involves rare discoveries that can't be replicated at will.

One of the most important aspects of validating scientific conclusions is the ability to replicate results. However, archeological discoveries tend to be rare—and they only get more rare the farther back in time you go. Sometimes there's very little to compare a site to, which makes it even more challenging to draw conclusions.

The act of discovery can destroy most—if not all—of the key evidence.

For example, if you have to dig to uncover a site, as you're digging it up, you're effectively destroying all of the evidence of the site's stratigraphic positioning. Archeologists are trained in how to conduct this sort of excavation and carefully document every stage to preserve as much information about the original condition of the site as possible.

Knowledge of key elements is entirely dependent upon the observers who were present.

However, no matter how careful someone is, it would be impossible for anyone to record absolutely everything about a site as it existed in its original condition. Any small detail that is overlooked could potentially have huge implications for the conclusions that are later drawn about a particular site, but once it's been excavated that evidence is destroyed forever.

Interpreting an archeological site is never simple.

On any given day you could take any two top archeologists, plop them down at the same site and ask them their conclusions and get two different answers. Coming to a consensus is often difficult, and made more so by the fact that the only data available is that which was collected by those who were on the site at the time of the dig.

There are incentives for people to lie.

We'd like to think that we can always depend upon a scientist to be objective and truthful in their conclusions, but the reality is that scientists are subject to the same conscious and unconscious biases—and the same temptations—as the rest of us.

While there are a few archeologists who manage to make multiple important discoveries in their careers, the reality is that many brilliant and talented archeologists toil for their entire careers without ever personally finding anything of import. Beyond just earning the respect of their peers, making an important discovery means more money for research and excavations, better positions at better universities, and even potentially book and media opportunities. It's not hard to understand why someone might leave out information that might contradict the importance of their find or might embellish slightly to make something seem more important than it was.

And I'd argue that there is the opposite pressure, as well—to gloss over or overlook anomalous details that contradict earlier findings and conclusions in the field. As with anything else, archeology has a deeply enmeshed established narrative regarding the history of humanity—and those who choose to challenge the status quo are often met with hostility and derision. They can be labeled as pseudoscientists, if not outright con artists. And for people who are slapped with that label, the fallout can be personally and professionally devastating. The research money and cushy positions dry up. They become outsiders.

Flaws With The Peer Review System

Which brings us to another epistemological challenge that impacts not just archeology, but the practice of science as a whole. Which is that there are flaws in the peer review system that can make it difficult for particularly challenging or unpopular ideas to get the same consideration as ideas that more closely align with the established narrative.

If you're unfamiliar, peer review is the system used to assess the quality of a manuscript before it is published in a scientific journal. Independent researchers in the relevant research area assess submitted manuscripts for originality, validity, and significance to help editors determine whether a manuscript should be published in their journal.

Basically, it's a way for other experts in the field to look at how the research being presented was conducted, to assess the conclusions that were drawn, and to give an opinion on how legit it is. To get published, scientists basically need to have a group of their peers look over their work and say, "This looks right to me." And peer review isn't just used to determine which articles are published, but to determine who receives scientific grants, which projects are funded, and which scientists are hired and promoted.

And that makes sense, right? We need some sort of system of checks-and-balances in science. We need a way to sort out the experiments that were poorly constructed and the conclusions that were unsound. We need a way of coming to consensus.

And don't get me wrong, for all its flaws, the peer review system is the best answer we have for that, by far. It's like democracy—it's a messy, imperfect system, but we have yet to come up with anything better. So please don't hear what I'm about to say as an attack on the peer review system or a suggestion that it should be abandoned. But I do think it's important to understand what its weaknesses are, and how they can become particularly evident in the case of scientific findings that radically alter our existing beliefs.

So what are those limitations?

Reviewers are usually untrained, unpaid, and overworked.

Now granted, "untrained" means something very different when you're talking about a scientist with one or more terminal degrees. The people who are conducting peer review are highly trained in both their field and in scientific methodologies as a whole.

However, they aren't generally trained on the peer review process itself. And, as we'll see, there are certain mistakes and biases that can impact what studies get published and which ones don't—which in turn impacts which scientific findings are given credibility and which ones are dismissed. So a little training on how to check yourself for those biases throughout the process seems warranted—but it's not something that most peer reviewers ever receive.

And with 2.5 million peer-reviewed articles being published annually—a number which does *not* take into account all the papers that are rejected—it's difficult to find enough people to do the work. Add in the fact that they are generally unpaid for their labor, and the result is a pool of peer reviewers who are well-meaning, but almost certainly overworked.

There is evidence that reviewers aren't always consistent.

In a 1982 study, two researchers selected 12 articles that were already accepted by well-respected scientific journals and switched the names of the authors and academic institutions to fake ones. They then resubmitted the exact same articles to the same journals that had already accepted them in the previous 18 to 32 months. Surprisingly, only three of the papers were identified by the editors and reviewers of these journals as being an article they'd already published. And of the nine that continued on, all but one was turned down with 89% of the reviewers recommending rejection.

Obviously, this lack of consistency suggests that there are factors beyond just the black-and-white scientific merits of an article that contribute to whether or not it's published.

There is evidence that the most innovative and impactful scientific ideas are more likely to be rejected.

A 2015 study tracked the popularity of rejected and accepted manuscripts at three top medical journals and found that, while the editors and reviewers generally made good decisions regarding which manuscripts to publish and which to reject, some of the most highly cited articles were the most likely to be rejected.

They started with 1,008 manuscripts, 808 (or about 80%) of which were eventually published. However, among those that were rejected were *all* of the 14 most cited articles. This suggests that while the peer review system is great for raising the overall average quality of articles that are published, it isn't very good at recognizing and promoting the most important and impactful research. Which makes sense—if you want to smooth out the mean of any data set, you first need to eliminate the outliers.

But sometimes, the outlier is the data point that changes everything. What then?

Putting It All Together

The takeaway here is this—the peer review system is a necessary and well-proven process for maintaining an overall level of quality and scientific rigor with regard to the scientific articles that are published. However, the system is not without its flaws, and has a general bias toward the status quo and against more innovative or radical ideas.

So putting it all together, what does all of this mean for us?

The Dunning-Kruger Effect

What it *doesn't* mean—and I want to be very clear about this—is that we should just throw everything out the window, or assume that we are as good as or better than archeologists at interpreting data in their field. We absolutely are not.

Right now, we're feeling smug and full of facts. We know about the challenges of dating archeological sites. We know that most archeologists only ever get to assess second or third-hand data. We understand the biases inherent to the peer review process. We're feeling very smart—and who doesn't love feeling smart?

But everything that we've talked about in the last few minutes wouldn't even fill one chapter of a 101 textbook. It's not a drop in a bucket, it's a drop in the ocean compared to the knowledge of people who have spent decades of their lives dedicated to studying these things.

When we find ourselves in this place, the Dunning-Kruger Effect is a good touchstone to keep us grounded. If you're unfamiliar, the Dunning-Kruger Effect is an extremely common form of cognitive bias where people who have very limited knowledge of something greatly overestimate how much they know about that thing.

It's basically a case of not knowing what you don't know. If you only know a little bit about something, you can easily make the mistake of assuming that what you know is basically all there is to know. And interestingly, once people start to learn more about a subject, and become more aware of its true scope and complexity, they very quickly go from overestimating their knowledge to realizing that they know next to nothing.

So basically, if you haven't studied something deeply enough to have been truly humbled by it, you probably know much, much less about it than you think.

Is there a conspiracy in academia to cover up the true origins of human civilization?

I also want to be clear that, although I do think there is more than enough evidence for us to question the established narrative about the age and origins of human civilization—I absolutely do *not* believe that there is a conspiracy within academia to cover any of this up.

Because I don't think that there needs to be. When I look at all of these things together, I don't see a plot. I don't see sinister machinations. What I see is our humanity, in all its nobility and absurdity.

We want to belong, so we shift and mold our beliefs to align with the whole. Our need to know who we are is so profound that we tell ourselves that we already do just to ease the existential friction. We mistake the limits of our knowledge for the limits of what is. We strive. We falter. We're wrong. We rise and try again.

It's just who we are.

What is the difference between science and pseudoscience?

And finally, I want to take a minute to talk about a word that gets thrown around a lot when we're talking about scientific ideas that threaten the established narrative—and that word is *pseudoscience*. And for any scientist who hopes to maintain the credibility necessary to have a career, much less get money to fund their research, it's the ultimate kiss of death.

But what is pseudoscience?

Pseudoscience is a collection of beliefs or practices mistakenly regarded as being based on the scientific method.

So if a person is using the scientific method to define a question, make predictions, gather and analyze data, and then draw conclusions—that's science. If a person is *not* using the scientific method—that's not science. And if someone is doing research and drawing conclusions in a way that they think is in alignment with the scientific method but actually isn't—that's pseudoscience.

Pretty straightforward, right?

And I come back to this idea again and again, because it's in this distinction that we can begin to get a solid foothold to help us evaluate whether or not ideas that have been relegated to the realm of pseudoscience may have actual merit. And best of all, we don't need to be experts in anything to do so.

It's as simple as this: is this idea being called pseudoscience because the approach is unscientific or because the conclusions don't conform to the established narrative?

If it's unscientific that should be easy to demonstrate because either the scientific method wasn't used or it wasn't used correctly. Maybe a variable wasn't accounted for. Maybe there was a flaw in the way that the data was collected. There are lots of possibilities, and any true expert in a field that is leveling an accusation of pseudoscience against a colleague should be able to clearly articulate their basis for saying so.

But what's been kind of astonishing to me as I've pursued this line of questioning is how often accusations of pseudoscience mention none of those things and focus instead on the fact that the conclusions being drawn by this person are "impossible."

But how many impossible things has science proven to be possible? "Impossible" is a meaningless term for a species as young as our own. What could we possibly know about what's impossible?

Impossible in this context means that it doesn't conform to the conclusions drawn based on the existing data set. That doesn't mean it's wrong. It just means that it doesn't fit. And in science we don't shape our data to conform to the conclusions—we shape our conclusions based on the data. The fact that a data point doesn't fit isn't grounds to throw it out—it's a hint that there may be a piece of the puzzle that we're missing.

And so while we may not be archeologists or scientists, I would argue that we are at least able to determine whether a set of ideas that is labeled as pseudoscience has been labeled so fairly. That doesn't mean that it's then necessarily true, only that we should wait to reject it until we're able to do so on its scientific merits.

And that, at least, gives us a place to start.

Where do we go from here?

So, as we conclude this introduction that became an entire chapter, I feel like we have a solid foundation to go forward from here and begin to explore some of the mysteries surrounding the dawn of human civilization.

We recognize the necessity of reevaluating the story of humanity through the lens of the UFO phenomenon. We understand some of the challenges that exist in creating a reliable and consistent narrative around events of the distant past. We know that although the peer review system helps ensure that a high level of scientific rigor and excellence is maintained in the work that is published, it also has a tendency to reject the work that is most important and impactful. And finally, we have a simple framework to help us assess whether radical and innovative ideas are being dismissed as pseudoscience based on their merits or if other biases may be at work.

And that's where we'll pick up next time, as we sift through the sands of time looking for answers, and begin to explore some of the most astonishing evidence that suggests that the history of human civilization on this planet may be far older and more dazzling than we ever imagined.

9 A Rational Approach To Ancient Aliens [Part 2]: The Secret History Of Human Civilization

In this chapter, we'll dive into part 2 of our discussion which aims to put forth a rational approach to the question of whether or not the UFO phenomenon may have interacted, and perhaps even intervened, with humans in the distant past—and if so what impact this interaction may have had on our species and the development of human civilization.

This idea, referred to as the ancient astronaut theory, and popularized by the show *Ancient Aliens*, is one that emerges again and again as you begin to study ufology. The evidence that the UFO phenomenon has played a clear, albeit baffling, role in human history is considerable—something that we'll get into more in book two—and is woven throughout our stories and art around the world spanning back as far into our past as these things appear.

There's something undeniably different about humanity. As far as we're aware, despite billions of years of evolution on this planet, humans are the first species to appear with the capacity for what we define as "intelligence". We're seemingly the first species to create art, to develop complex language, to develop agriculture, and to build cities.

We're the first species to not just wonder about who we are in the grand scheme of the cosmos, but to develop complex frameworks and methodologies to begin to pursue the truth behind these mysteries ourselves. I mean, we went to the fucking moon.

And yet, our nearest cousins the chimpanzees, highly intelligent though they may be, are only fractionally closer to writing the Declaration of

Independence than your average dog. Despite sharing more than 98% of our DNA, the differences between us are undeniable and profound.

And the reality is that we have no idea why we are so different. The unique quality of our humanity is self-evident, but where did it come from? Are we just very smart apes who through trial-and-error, and no small amount of luck, over millions of years happened upon the path to math, philosophy, art, agriculture, and government?

Or is there perhaps more to it than that? Could we have been given a boost—some form of guidance or revelation from a more advanced intelligence? The mysteries of our past certainly seem to raise this intriguing possibility.

So let's take some time to dive into the evidence surrounding the emergence of one of the most unique and astonishing of humanity's achievements—the development of high civilization—to see what clues there might be about this profound chapter of our history, and what role, if any, the UFO phenomenon may have played in its emergence.

What Is Civilization?

Before we go any further, we need to stop for a minute to get some clarity around what civilization actually is. After all, if we're going to look for clues about human civilization in the distant past, we need to know what we're looking for, right?

And the concept of civilization is one that may seem obvious, but upon closer inspection is anything but. We use that word in a lot of different ways in our everyday lives. For some civilization is running water. For others, a place needs to have at least one Michelin Star restaurant to qualify. And when we look at more technical definitions of civilization in an attempt to understand and trace its emergence, things get even more complicated.

There are a lot of layers and a lot of history around the very idea of civilization that we need to unpack in order for us to take a more informed and nuanced look at the evidence and begin to understand what secrets may lie in our distant past.

What Is The Difference Between Civilization and Culture?

The word "civilization" brings to mind so many of the things that make humans unique among other animals on this planet—things like art, language, knowledge, technology, and government. But how would we define it exactly?

An important distinction to make here is the difference between civilization and culture. Culture is the sum total of ways of living built up by human beings and transmitted from one generation to another. Culture is a complex entity that's made up of a confluence of disparate parts and can include things such as art, customs, morals, beliefs, religion, food, and even laws.

Civilization, on the other hand, is an advanced state of human society, in which a high level of culture, science, industry, and government has been reached. A civilization is made up of and springs from culture, but is also associated with a number of social, political, and economic characteristics such as urban development, social stratification, unique art and architecture, a form of detailed writing, complex division of labor, codified laws and administration, and more.

So art? Culture.

Museums? Civilization.

Food? Culture.

Culinary School? Civilization.

A group of people can have and enforce laws as part of their culture, but in a civilization, those laws are codified and enforced by specialist groups assembled for that purpose like legislative bodies and police forces.

So this distinction gives us a decent place to start talking about how to truly define what a civilization is. Our culture emerges from our humanity—our sense of beauty, our sense of order, our sense of right and wrong. And culture can exist without civilization. We see this in humanity's past where we have examples of cave art dating back at least 40,000 years.

But civilization, on the other hand, emerges from complexity—complex ideas, complex ways of meeting needs, complex social structures.

And yet still this definition feels a little soft. Where is that line between the simple and the complex? How do we definitively say when a group of people have crossed that threshold, and how would we recognize the signs of that transition in the distant past?

Writing As a Hallmark Of Civilization

Because of the complexities of defining what exactly a civilization is, many experts have settled on written language as the defining characteristic. And this makes sense in a lot of ways.

First of all, writing gives us something concrete and "discoverable" by which to define civilization. Things like complex social structures and codified laws can be inferred about a group of people based on the evidence they leave behind, but without a written language to decipher, it is difficult—if not impossible—for us to know much about them.

Written language itself also implies a level of social complexity that we associate with civilization. Its very existence implies the need for complex information to be conveyed to a large group of people. It implies some kind of formal (or at least structured) education. It implies a permanence, or perhaps more accurately a *persistence*, of a group's culture and a sense of identity.

For these reasons, written language is an intuitive benchmark, and if you have to draw a line in the sand somewhere to clearly define what civilization is, written language feels like a good place to do it.

Challenges With Using Written Language To Define Civilization

However, there are some issues with this approach, as well.

I'd argue that for written language to be used as a reliable benchmark for human civilization, then, at minimum, two things need to be true of the examples of written language that are left behind by that civilization—they must be durable and recognizable.

Let's take a minute to talk about what I mean by that.

So first of all, for us to find this writing, it has to still exist somewhere, which means that it has to be durable. It has to survive long enough for us to find it.

However, more often than not, written language isn't very durable— or rather, the materials that are used for written language aren't very durable. Organic materials like paper and papyrus, for example, eventually decay.

They're also particularly vulnerable to damage from everything from fires to floods to simple humidity.

The Library of Alexandria comes to mind. Believed to have been built sometime around 250 B.C., the Library of Alexandria in Egypt was perhaps the largest and most significant library of the ancient world. At its height it was said to have housed anywhere from 40,000 to 400,000 scrolls.

Despite a common misconception that the library was burned down entirely in one catastrophic incident, it was actually partially burned three different times over a few hundred years during various wars and conflicts before falling into decline due to lack of funding and eventually disappearing entirely sometime in the late third century. (And this is why you've always got to vote to fund your libraries, people. Do your part.)

And though the story is different from the story of the Library of Alexandria that is typically told, it doesn't really matter—the impact is the same. For hundreds of years, an astonishing wealth of human knowledge was meticulously stored and protected in the Library of Alexandria, and yet nothing of it now remains. Eventually disasters, wars, and cultural atrophy devoured it all.

Interestingly, even in our current day we are confronted with the challenges of preserving and safeguarding records for future generations. This is because more and more knowledge and information is being created and stored exclusively in a digital format. And although technology makes it much easier to duplicate, back up, and share this information, the technology we use to access it, as well as the formats in which it is stored, are constantly changing.

There are those who think that we're not doing enough to preserve our digital records, and that we are at serious risk of creating a black hole in our history. Think about it this way—how much information was stored on floppy disks and CDs and other relatively short-lived data storage mediums? And how far forward into the future would you need to go for that data to become virtually irretrievable? A couple hundred years perhaps? Maybe less? For anyone who is unlikely to invest any energy is this sort of thing—which I would argue is basically everyone—those things are effectively irretrievable now.

So what can we use to preserve written language in a way that can withstand that test of time?

The most obvious answer is with pottery or clay tablets—or if you really want to get serious you could carve it into stone. And although we've undoubtedly lost immense amounts of human history and knowledge that was stored on less durable materials—luckily, our ancestors did use these sorts of materials, as well.

The very earliest examples of writing that we have are on clay tablets. They date back to around 3500 B.C. in what was once Mesopotamia and is now modern-day Iraq. This is where it is believed that written language was first developed. We have examples of early pictographic writing starting about 3500 B.C. that slowly transitioned into what we recognize today as cuneiform.

Which brings us to our next point, which is that for written language to be an effective benchmark for the emergence of human civilization, we need to be able to recognize it as writing.

But a written language is just a set of symbols that convey meaning. And there are many different kinds of languages. In English, each character represents a sound. In some languages, each character represents a syllable. And in others, each character is a word or an idea. Some of these characters can be relatively simple like in the Latin alphabet, or they can be complex pictographs where minor variations in the character can convey subtle layers of meaning.

Chinese consists not only of characters representing meanings but also of secondary characters based on sound similarity for representing meanings that were difficult to picture. It therefore relies upon both word-based and sound-based principles.

But, whatever the language, if we don't understand the meaning that these characters are meant to convey, how would we even know for sure that what we are looking at is truly a written language? What would separate random characters and pictographs from any other markings one might find on walls or tablets in ancient archeological sites?

Because preserving written language over thousands of years presents so many challenges, and because we may not even recognize it as written language if we do find it, I think it's important to recognize the very real possibility that we could easily miss written language in the archeological record.

Which, I would argue, means that written language can't be a viable benchmark for identifying human civilization. And so what we're left with is a question for which we still don't have a great answer—which is *what is civilization?*

The Legacy of V. Gordon Childe

I want to take a minute to talk about the ideas of one man that have perhaps shaped the current accepted narrative of the emergence of human civilization more than any other—and that man is V. Gordon Childe.

Childe was a famed Australian archaeologist in the early 20th century who specialized in the study of European prehistory. He spent most of his life in the United Kingdom, working first as an academic for the University of Edinburgh and then the Institute of Archaeology in London.

Childe was known as a great synthesizer of archeology and wrote twenty-six books during his career. He initially made his mark as an early proponent of culture-historical archaeology, which is an archaeological theory that emphasizes defining historical societies into distinct ethnic and cultural groupings according to the art and material that their culture produced.

However, later in his career, he became a staunch proponent of Marxist archaeology, which is a theory that interprets archaeological information within the framework of Marxism. Now, whatever your opinion on Marxism, I think we can all at least agree on the idea that interpreting the lives and cultures of ancient people entirely through the lens of a modern socioeconomic theory is, at best, a dodgy approach.

Personal Bias In Assessing Ancient Civilization

But this is also something that we, as humans, do a lot. We tend to apply the blueprint of our current understanding to unfamiliar cultures and concepts, assuming that they are more like us than they actually are.

I spent a few years traveling and living abroad, and until I had that experience, I don't think I truly understood how much that's true. Every culture is deeply coded with nuanced layers of meaning threaded throughout each and every interaction. We don't notice it when we're embedded within our own culture, because that culture is the context in which we live our lives. It's as natural to us as breathing.

But when we find ourselves in a foreign culture, suddenly all of that comes sharply into focus. We can be easily confounded by the simplest things. Do you wait to be seated or seat yourself? Do you tip? Do you shake hands when you meet? Do you hug? Do you bow? Or maybe you go for the European cheek kiss—but one cheek or two? Surprise! You're in Amsterdam and the answer is three.

I lived in Amsterdam for a couple of years, and I found myself in these situations all the time. And it says a lot that this was true even in a place like Amsterdam—a Western, democratic country with a shared Eurocentric historical perspective and whose citizens are basically all conversational or better with English. But despite those many similarities, the Dutch perspective is so entirely different from the American one that I was constantly caught off guard by something that seemingly should have been obvious to me, but wasn't.

Here's a quick example of what I mean:

A popular tourist attraction in Amsterdam during the time that I lived there was a giant sign made out of enormous block letters that said "IAmsterdam" with the "I" and the "am" in red and the rest in white, so it effectively said "I am Amsterdam". These signs were all around the city from the airport to the world famous Museumplein, and people would climb up on the letters and take pictures for the 'Gram.

However, in 2018 it was announced that the popular installations would be removed. The decision was made by majority vote by the Amsterdam City Council in response to a motion filed that said that the sign was "too individualistic" and didn't promote "solidarity and diversity".

Now, to most Americans, this is a baffling argument. We are a nation founded on the almost worshipful notion of rugged individualism. We see individualism as the ultimate virtue, not a societal vice that needs to be overcome. And while we also believe in the ideals of solidarity and diversity, we see it as solidarity among self-defined individuals. We see individualism as the cornerstone of diversity, not the enemy of it.

But to the Dutch, individualism is at the root of what they see as an increasingly sick and unbalanced western culture. It is antithetical to their collectivist beliefs that place the needs of the whole above the needs of the individual. They see aggressive individualism as the root cause of much of the injustice and inequality in the world.

And to be honest, even as I'm saying this, I'm not 100% confident that I'm properly conveying their point of view on all of this because I don't know that I totally understand it—the idea is just so fundamentally foreign.

The question of who's right and who's wrong doesn't matter, and is probably asinine. The point is that even cultures that exist at the same time period and which have significant amounts of overlap in their shared experience can be very different from one another and can operate with entirely different sets of beliefs and motivations that would likely not be immediately obvious to outsiders.

Which is why I'd argue that it is almost impossible for us to easily understand how different a culture that existed 2000 years ago is from our own. And when you go further back into time—to 5000 years ago, 10,000 years ago, and beyond—those differences inevitably become more profound. And so, to conduct effective archeological inquiry we need to always be cognizant of this fact and do our best not to place the blueprint of our experience on top of things that we don't understand.

So—getting back to V. Gordon Childe, the takeaway is that his ideas changed a lot throughout his career. And while he was well-respected and made important contributions to archeology, many of his ideas have come into question or have been outright disproven since.

However, despite this, those same ideas still form the basis for much of the mainstream narrative around the emergence of human civilization to this day. As we talk through some of these ideas, they'll likely be familiar to you.

Savagery, Barbarism & Civilization

So, first of all, Childe divided preindustrial societies into three main categories: savagery, barbarism, and civilization. And these aren't just categories, but represent a hierarchical progression from least to most advanced. According to Childe, "savages" came first—or to use a more accurate definition, these were hunter-gatherers. These people lived off of wild foods that they obtained by hunting, fishing, gathering foraging, and scavenging.

Next came "barbarians". These groups of early humans might still hunt and gather to some extent, but they also began to supplement their food supply by cultivating plants and/or herding and breeding animals.

Cultivation and herding necessitated that small groups of people band together to protect small patches of land being used for that purpose. People began to weed, water, irrigate, and even engaged in selective breeding in order to get more yield out of their land. The tasks around cultivating this land became more complex, but they also lead to more abundant outcomes—AKA more food. As a result, larger and larger communities arose around this land.

And then finally, once these communities reached a certain size and level of complexity, Childe defined this as "civilization".

Problems With Childe's Model

Pretty straightforward, right? And it's probably not too different from what you learned in school. However, there are some serious issues with Childe's models that are worth discussing.

First of all, Childe himself admitted that it was difficult to clearly define where the line between "barbarism" and "civilization" actually lies. He generally equated civilization with urbanism and "city life", but recognized that that was also a difficult thing to define, so he tended to agree with the view that written language was the best benchmark for identifying true civilization. And we've already discussed how that approach, while convenient, is inherently problematic.

The second issue is that the language that he chose to describe these different stages of pre-industrial human societies is coded with a cultural bias that can lead to incorrect conclusions.

For example, a savage is defined as "a brutal or vicious person" or "a member of a people regarded as primitive and uncivilized". On the other hand, a hunter-gatherer is "a member of a nomadic people who live chiefly by hunting and fishing, and harvesting wild food." Do you hear the difference?

And I want to be very clear that what we're talking about here isn't political correctness. This is about accuracy in word choice and guarding against unconscious bias in our understanding of ancient people. And we can easily see in this example how the use of the word "savages" to describe people who are hunter-gatherers layers unnecessary and unintended meaning onto the concept that can obscure our understanding.

We assume that because a group of people were hunter-gatherers that they were primitive, brutal, and uncivilized. We assume that high culture and advanced knowledge can only exist within a "civilized" society, which is a concept that we've arbitrarily tied to "city-life" and "written language" even though we have trouble clearly defining and identifying either of those things—and even though we can't clearly articulate a reason why those things should be the ultimate defining characteristics of civilization and advanced human achievement.

And although there is abundant evidence—which we'll get to in just a minute—to show that these ideas are flawed and don't accurately represent the emergence of human civilization as it actually occurred, they still persist both within the cultural zeitgeist and within archeology itself.

The Story Of Human Civilization (According To Mainstream Academia)

Alright, so now that we've laid the foundation, let's talk about the story of human civilization according to mainstream academia.

We think that modern humans may have first appeared somewhere around 300,000 to 350,000 years ago. I say "may have" because the truth is that we are far from certain about where and when homo sapiens first came on the scene.

Up until very recently, the earliest examples of homo sapien bones that we had discovered were from about 195,000 years ago in modern-day Ethiopia. It was thought that this was the metaphorical Garden of Eden, where humans had first evolved before spreading out into the rest of the world.

However, in 2017 a new discovery was announced. At a site near Morocco's coast, near the city of Marrakesh, bones belonging to modern homo sapiens were found dating back 300,000 to 350,000 years. Not only did this discovery nearly double the history of the human race, but it calls the idea that humans first evolved in Eastern Africa seriously into question.

Regardless of when humans first appeared, it is generally agreed that around 70,000 years ago our species underwent a cognitive revolution. We began to live in larger communities, we engaged in trading with other groups, we developed speech, and even creative art. We're not sure why this happened, but we've been able to identify these changes in the archeological record.

And then, about 12,000 years ago, humanity encountered two important turning points. The first was the end of the Younger Dryas, which was basically a mini Ice Age that lasted for about 1,200 years.

At the onset of the Younger Dryas there was a massive, worldwide extinction of large mammals weighing over 40 kg (or 88 pounds). It is estimated that 82% of these animals disappeared in North America, 74% in South America, 71% in Australasia, 59% in Europe, 52% in Asia, and 16% in Sub-Saharan Africa. This mass extinction event marked the demise of the mammoths, as well as the disappearance of horses in North America, and other species, including bison, deer, and moose suffered massive population losses.

Fossil evidence suggests the disappearances were very sudden, and so whatever caused that sudden die off inevitably made things mighty uncomfortable for our ancestors, as well—particularly in the Northern Hemisphere where the impact appears to have been highest.

However, despite whatever hardships they endured, humans emerged from the end of the Younger Dryas having managed to not suffer the same fate of most other large mammals. And it was at this point that humanity reached another turning point—the development of agriculture.

This transition to an agricultural society is also known as the Neolithic Revolution—a name that was coined by our old friend V. Gordon Childe. According to mainstream archeology, the Neolithic Revolution began around 12,000 years ago in the fertile crescent, also known as Mesopotamia, surrounding the Tigris and Euphrates Rivers in what is now modern day Iraq.

This Neolithic Revolution marked the transition in human history from small, nomadic bands of hunter-gatherers to larger, agricultural settlements and early civilization. It was a radical and important period of change in which humans began cultivating plants, breeding animals for food and forming permanent settlements.

This wasn't a transition that happened over night. And it would take 6 or 7 thousand more years for these agricultural societies to reach a level of complexity that would trigger the next big transition for humans—the urban revolution.

The cumulative growth of technology and the increasing availability of food surpluses allowed people to begin to live together in larger groups with a more complex social structure. Specialists emerged, allowing for significant advances in the arts and sciences. Surpluses were traded for other capital to fund public works. The first examples of written language emerged.

And, just like that, you have civilization. And almost immediately, these civilizations advanced to the level that they could begin to undertake megalithic building projects.

Which, if you ask me, sounds a little fishy.

The Pieces That Don't Fit

But that's more or less the story of human civilization as told by mainstream archeology today. However, over the past few decades there have been more and more discoveries that seem to not just contradict this narrative, but suggest that we may need to go back to the drawing board and rethink the timeline of the emergence of civilization entirely. Yet, strikingly, these discoveries have done little to dislodge this narrative among most academics.

And no archeological site better exemplifies the level of dissonance between the accepted narrative and what the evidence actually seems to suggest than Gobekli Tepe.

Gobekli Tepe

I want to start with Gobekli Tepe, because this is the archeological site that blows up our established narrative. It is irrefutable proof that our timeline and assumptions about early human civilization are wrong. And it poses new questions about our origins that are not easily answered.

In southeastern Turkey, just north of the Syrian border, atop a low mountain sits Gobekli Tepe. Discovered in 1994 by German archeologist Klaus Schmidt, with excavation beginning in 1995, this remarkable site challenges everything we believed to be true about the emergence of human civilization.

The site comprises a number of large circular structures supported by massive t-shaped, stone pillars—which are the world's oldest known megaliths and are estimated to weigh as much as 20 tons each. Many of these limestone pillars are richly decorated with abstract anthropomorphic details, clothing, and reliefs of wild animals.

There are many remarkable things about Gobekli Tepe—its complex geometry, its intricate stone carvings, and its towering megaliths. However, the most remarkable thing about this incredible site is its age. Using multiple dating methods, it is believed that the earliest parts of Gobekli Tepe were built around 9600 B.C.—or nearly 12,000 years ago. And yes, this is the date that is accepted by mainstream archeologists.

The reason that Gobekli Tepe is so well preserved, and why we're so confident about its astonishing age is that, for reasons that remain a mystery, the entire complex was meticulously buried sometime around 8000 B.C., and remained that way until its discovery in the late 20th century.

It's hard to know what, if anything, would have remained of this site if it hadn't been preserved in this way, emphasizing how rare and important this glimpse into our past really is.

What Can We Learn From Gobekli Tepe

Well first of all, we now know that, at least in what is modern-day Turkey, humans were building megalithic structures at the end of the last ice age, which is a full 5000 years before the first megaliths were believed to have been built.

To put that into perspective, the length of time between when the first megalithic structures were believed to have been built and when Gobekli Tepe was actually built is roughly the same period of time between our present day and the early dynastic period in Egypt, hundreds of years before the pyramids were believed to have been built. This is stunning for a number of reasons.

It completely obliterates our previous timeline of the emergence of human civilization.

According to the existing narrative, 12,000 years ago humans were just beginning to experiment with agriculture. They still lived in small groups and were primarily hunter-gatherers. However, carving and raising giant megaliths like the ones found at Gobekli Tepe implies a level of sophistication that blows this story out of the water. To build something like that takes time, planning, and enormous resources—how could that be accomplished by small groups of hunter-gatherers who would, by necessity, need to exert most of their energy to get food?

To be fair, we actually don't know if the builders of Gobekli Tepe were truly nomadic, hunter-gatherers. It's been speculated that there could have been a permanent settlement near the site, though no evidence of that has ever been found. It could be that their homes were far less elaborate and made of largely organic materials—in which case it's unlikely that any real evidence of their existence remains. Or it's also been speculated that the settlement could have once stood where the nearby city of Urfa is now, which also would make it unlikely that we'd ever find traces of it.

However, it's also possible that there wasn't a permanent settlement at Gobekli Tepe at all. Perhaps Gobekli Tepe was a gathering place for various groups and tribes to come together for trading or religious celebrations. If that's the case, it only further muddies the waters of the established narrative, because the level of knowledge and scale of cooperation necessary to take on megalithic building projects wasn't supposed to have been developed until 5000 years later—and even then, it was only supposed to have been possible within the context of a "civilized", urban environment.

Honestly, either possibility is mind-blowing.

It may contain evidence of written language.

Carved into one of the largest megalithic pillars is what looks very much like letters. Specifically, it looks like a capital "C", followed by a capital "H", followed by a backwards capital "C". Now, obviously, these aren't literally C's and H's. However, comparisons have been made between these markings and Anatolian hieroglyphs that were used in the area, albeit 8000 years later, in the second millennia B.C.—and the similarities are striking.

Could it really be possible that written language developed thousands of years earlier than we previously thought? And could the written language used by the people who built Gobekli Tepe have eventually been the basis for Anatolian hieroglyphs?

Frustratingly, the answer is that we simply don't know. As compelling as the evidence for written language is at Gobekli Tepe, there's not enough to draw any definite conclusions—which only further emphasizes the point that identifying written language in our distant past is far more complicated than we might initially assume. The characters could be written language or it could be decoration—without context, how can you make that discernment?

And whether or not these markings are written language, we're still left with a conundrum. If it is written language, then we have to push back the advent of the written word by at least 7,000 years. And if it isn't written language, then we have to grapple with the fact that the builders of Gobekli Tepe somehow reached a level of social complexity to produce such a massive and mathematically precise megalithic construction without it—meaning that written language is no longer a viable benchmark for the emergence of civilization.

Either way, the accepted paradigm is shattered beyond repair.

The development of agriculture may not have been the driving force behind the Neolithic Revolution.

Because here's the thing—the agricultural revolution is dated to about 10,000 B.C., but making the shift to fully agricultural societies took thousands of years. It didn't happen overnight.

Written language, complex society, and advanced human knowledge may have developed long before cities. There may have been some other driving force besides the agricultural revolution that caused people to abandon their hunter-gatherer lifestyle for larger and more permanent settlements. The emergence of agriculture could have been a result of this transition, as opposed to the cause of it.

Whatever the case, Gobekli Tepe stands apart as the single most significant archeological find in human history, and its mysteries challenge virtually every belief that we have about the dawn of human civilization.

The Sphinx

In light of the revelations of Gobekli Tepe, it seems reasonable that we would go back and reexamine other ancient sites within the context of this new information. However, despite the fact that excavations began at Gobekli Tepe almost 3 decades ago, within most archaeological circles, the idea that megalithic structures may be older than we thought—or at the very least, that they could have been built on the foundations of much older sites—remains entirely taboo.

One of the most well-known examples of this involves what is perhaps the dopest of all ancient megalithic structures, The Great Sphinx.

In October 1991, at an annual meeting of the Geological Society of America, Dr. Robert Schoch first presented evidence that the origins of the Great Sphinx must date back to at least 7000 to 5000 B.C., if not earlier—which is thousands of years before the popularly accepted date of 2500 B.C. And in more recent years, he has pushed back what he believes to be the true age of the Sphinx to around 12,000 years ago.

Schoch's theory has been largely criticized and dismissed by mainstream archeologists and geologists, and his work has been slapped with the label of pseudoscience.

And as we talked about in the last chapter, we may not be archeologists or geologists, but we do have a fairly simple framework for determining whether or not something is being fairly labeled as pseudoscience. All we need to do is to look at the arguments for why something is being called pseudoscience. If the argument is that the methodology used to collect the data wasn't sound or that the data was interpreted incorrectly, then they might have a point, and we'd need to dig deeper to understand whether or not it was pseudoscience.

However, if the argument is simply that it's pseudoscience because the conclusions don't align with the current narrative, then we have a pretty clear case of academic bias. This doesn't mean that these conclusions are necessarily correct, it only means that the label of pseudoscience, at least as it's being argued, isn't fair.

So let's look at the evidence:

Mainstream Egyptologists contend that the Sphinx was built in approximately 2500 B.C. The reasoning for this basically boils down to the fact that it is surrounded by other structures that they have also dated to this time period.

The Great Sphinx is situated on the eastern edge of the Giza Plateau on the west bank of the Nile across from Cairo. Also located on this plateau is the Great Pyramid, attributed to the Fourth Dynasty pharaoh Khufu; the Second Pyramid, generally attributed to the pharaoh Khafre, who was possibly the son or brother of Khufu; and the third and smallest pyramid, which is attributed to the pharaoh Menkaure, who was possibly the son or grandson of Khufu.

The Proximity Argument

So based on this traditional thinking, all of the pyramids are dated to around the same time period around 2500 B.C. during and immediately after the reign of Khufu. And, they argue, the Great Sphinx is "associated" with these structures, so it must be around that same age, as well.

Now, I don't know about you, but that doesn't strike me as the strongest argument. I mean, sure—they're close to each other. But if you went to Rome right now, you'd find an assemblage of architectural marvels whose origins span hundreds, if not thousands, of years. And we'd be wrong in assuming that just because they are close to each other that they are the same age.

I'd argue that in any city, especially a city like Egypt or Rome, that has been continually occupied for thousands of years, that proximity simply isn't a viable method for dating structures.

The Pyramids May Not Be Correctly Dated

It's also worth noting that the dating of the pyramids themselves has, at times, been called into question. You probably already know that the three pyramids on the Giza plateau are positioned in such a way as to correlate to Orion's Belt, within the constellation of Orion. However, what's interesting is that, although their size and relative position are the same as the three stars of Orion's Belt, they don't align with the position of Orion's Belt in the sky in 2500 B.C.

This might seem like a minor detail, except that we know that the builders of the pyramids—like virtually all megalithic builders—had an incredibly advanced and detailed understanding of the movement of the stars and our place in the cosmos, and the the placement and orientation of these sites are almost always aligned to important astrological alignments, like the equinoxes. In fact, the pyramids themselves are perfectly aligned with the cardinal directions with a precision that humans wouldn't achieve again until the 18th century. So it's a little weird that they would be so precise in one respect, but so imprecise in another.

Which is why it's been suggested that perhaps the pyramids date back to a time when the pyramids would have been aligned with Orion's Belt. The only problem is that the last time that happened was 12,000 to 14,000 years ago.

Now granted, it's unlikely that the pyramids as they now stand were built that long ago because they would be much more heavily eroded. However, they could potentially be built on older sites that originated from that time period—or they could be memorializing that time period for reasons we don't yet understand. Either way, it's interesting.

Water Erosion On The Sphinx

And speaking of erosion, it was exactly that which caused Robert Schoch to call the age of the Sphinx into question.

Schoch's work is based on the startling realization that the considerable weathering shown on the oldest parts of the Sphinx show evidence of significant water erosion. However, you'd have to go back in time thousands of years before the Sphinx was allegedly built to find a time when there was abundant rainfall in the region.

Schoch says that he reached his conclusions "through a variety of independent means, such as correlating the nature of the weathering with the climatic history of the area, calculating the amount of rock eroded away on the surface and estimating how long this may have taken, and calibrating the depth of subsurface weathering around and below the Sphinx."

Some archaeologists and Egyptologists will concede that there is water weathering on the Sphinx, however they argue that heavy rains must have continued into the third century B.C. to account for this. Schoch counters this argument by pointing out that there are mud-brick tombs called mastabas built on the Saqqara Plateau only about a dozen kilometers south of Giza that are definitively and indisputably dated to 2800 B.C.—or 300 years *before* the Sphinx was allegedly carved. If there had been torrential rain capable of creating the deep weathering on the Sphinx in the area up through that time period and beyond, those bricks could not possibly have survived.

The Head Of The Sphinx May Have Been Recarved

There's also the question of whether or not the head of the Sphinx was recarved, something that mainstream archeologists vehemently deny. But—I gotta tell you guys—the first time I heard that theory my first thought was, "Oh, so *that's* why the Sphinx looks like that!" The proportions of the Sphinx, with its small head relative to its body always struck me as a bit off. However, as soon as I visualized it as a recarving of an original, larger head the whole thing just made sense in a way it didn't before. That's certainly not definitive proof of anything, but—I mean—just *look* at it.

Mainstream Egyptologists, however, contend that the head of the Sphinx is original and carved in the image of the pharaoh Khafre—the alleged builder of the Second Pyramid and the maybe-son or maybe-brother of Khufu. They argue that the bizarre proportions of the Sphinx are due to the fact that it was carved directly out of the limestone bedrock, with the body being carved first and then the head, and that they simply misjudged and ran out of room.

And here we see this paradox that conventional archeology demands that we accept without question, even as it emerges again and again—which is that we are to believe that the Egyptian people were both advanced enough to build the pyramids — an architectural feat that would be a major challenge to us, even today—*and* aligned them to the cardinal directions with a precision that humans wouldn't be able to achieve again for another 4000 years. And yet, they are somehow also the primitive bumblers who ran out of room to carve the Sphinx.

You simply can't have it both ways.

So if the head of the Sphinx was recarved, what did it look like before?

The most popular theory is the most obvious, which is that it used to have the head of a lion to match its body, and that perhaps this may have been a representation of the goddess Mehit in the form of a lioness. And here we once again run into the question of astrological alignment, because if the Sphinx was originally a lion, then might it have been carved during the Age of Leo?

It's impossible to know, but it's an intriguing theory. However, the fact that during the Age of Leo the Sphinx would have been facing the constellation of Leo on the vernal equinox makes this possibility even more compelling. The last Age of Leo lasted from 10,500 B.C. to 8000 B.C., which also aligns with Schoch's proposed age of the Sphinx at 12,000 years old. Once again, it's not proof, but it's an interesting possibility.

Arguments Against Shoch's Theory

I went looking for arguments against Schoch's dating of the Sphinx, because I truly wanted to understand why the academic community is so opposed to this idea. And I gotta tell you, I couldn't find much beyond what's been presented here so far.

There is one argument that the Sphinx is a lightly shaped yardang, or a rock outcropping shaped by wind-based erosion. Basically, the idea is that the Sphinx used to be a rock outcropping that looked kind of like a Sphinx and then they just carved it a little to make the resemblance more exact. This is a possibility, but given the way that the Sphinx is carved down into the bedrock, it feels unlikely that this was the case.

Other than that, the only real argument against Shoch's aging of the Sphinx was something that, when he first presented it in 1991 was a very fair argument—which was, if there was a civilization advanced enough to have created the Sphinx thousands of years before it was thought to have been built, then where was the other evidence of this civilization? Shouldn't there be *something*?

But just 4 years later, in 1995, excavations began at Gobekli Tepe. Here was the proof that humans were capable of massive and complex feats of engineering at the end of the last Ice Age.

So why, exactly, are mainstream Egyptologists so adamantly opposed to even considering the possibility that the Sphinx may be older than previously thought?

The Truth About Pseudoscience

In February of 1992, the American Association for the Advancement of Science hosted a heated debate on the subject between Dr. Robert Schoch and Dr. Mark Lehner of the University of Chicago. During this debate Lehner said, "You don't overthrow Egyptian history based on one phenomenon like a weathering profile. That is how pseudoscience is done, not real science."

And though this represents the consensus opinion of mainstream Egyptology, this statement is both telling and nonsensical. Egyptian history, like all history, is a story. It's a story compiled not of all that was, but of what little still remains, interpreted imperfectly by the dominant culture of the time period, and passed down and reinterpreted through the millenia. It's a messy game of telephone. History isn't a science. It's an echo. It's a shadow play.

So yes, when you find solid, observable data like, say, a weathering profile, and it doesn't match up with the story that we've told ourselves about what happened in the past we *absolutely should be willing to at least consider the possibility* that we need to throw that story out and start over. To say otherwise is to let the conclusions dictate how we interpret the data — and *that* is the very soul of pseudoscience.

And it is this exact type of pseudoscientific, nonsensical argument that is used to assail the reputations and ruin the careers of scientists who dare to question the established narrative—and it happens *all the time*. Once you learn to look for it, you'll see it everywhere.

So let's pour one out for the science homies—wherever and whenever they may be—who have put their careers and reputations on the line in order to follow the data to an unpopular and inconvenient conclusion.

Rewriting The History Of Human Civilization

Despite the vehement protestations of the academic community, I find it very hard to plausibly deny that between just these two archeological sites, we have evidence that a rewriting of history may be exactly what's in order.

But are just these two archeological sites enough to make that claim?

I'd argue that we only actually need one—Gobekli Tepe. The fact that it was carefully and meticulously buried 10,000 years ago means this site has been preserved in a way that we rarely see in megalithic sites—even those that are only a fraction of its age—giving us a literally unparalleled glimpse into humanity's distant past.

And what we've found looks *absolutely nothing* like what we would have expected and defies virtually every assumption we had about who these people were, what their lives were like, and the engineering feats of which they were capable. It's been almost 30 years since Gobekli Tepe was discovered. It's time to admit that when it comes to the emergence of human civilization, we had it all wrong.

On some level, mainstream archeology must know this. They don't dispute these dates, only the conclusions. But how can anyone deny that Gobekli Tepe nearly doubles the amount of time that there has been something like high civilization on this planet, however erratic and flickering that light may have been?

I think that, beyond the typical academic bias against paradigm-breaking ideas, there are a few other reasons for this resistance.

The first is that we still don't know what Gobekli Tepe was exactly. Most scholars seem to believe that it was probably a temple of some kind, used for religious rites and rituals—but we really don't know. And the completely anomalous nature of this site, seemingly standing alone in history without context, makes it even more difficult to make sense of.

Who were these people? And why did they build this place?

Mainstream archeology has chosen to simply set Gobekli Tepe down into the existing version of the story of what our ancestors were up to 12,000 years ago and will tell you that it was built by hunter-gatherers.

And to that, I simply say—how?

Could Hunter-Gatherers Have Built Gobekli Tepe?

First of all, the very nature of a hunter-gatherer lifestyle is that the vast majority of your time and energy is spent obtaining food. It's hard to understand how a massive, collaborative megalithic construction effort could be undertaken by people living that way.

For example, the massive megaliths at Gobekli Tepe are up to 5.5 meters (or 18 feet) tall and weigh up to 50 tons. It would take approximately a thousand men to move just one of these monsters, and though there are debates about how this was actually done, it was, undoubtedly, extremely slow, extremely difficult work. And there were 250 of these megaliths that needed to be quarried, moved to the location, and then raised.

So if you're going to get done with this project basically ever—you're going to need thousands of strong adults, who otherwise would have been hunting, to spend a *lot* of time very slowly hauling big rocks around. And meanwhile, there must be an equally massive effort to gather enough food to support this large assemblage of people. Dragging rocks is hungry work. And, conceivably, this could deplete natural resources in the surrounding area over a long enough time period, forcing them to go further and further afield to find food. None of this is ideal.

Now granted, there is the argument to be made that this is actually exactly what happened and that, instead of the rise of agriculture creating the conditions necessary for the building of megaliths as we'd previously assumed, that, in fact, the building of megaliths lead to the conditions that necessitated the development of agriculture. And I will admit that this is definitely a possibility.

However, I'd argue that it's not just time, manpower, and resources that would have been potential barriers to hunter-gathers becoming megalithic builders—but something deeper.

Many of the pillars and megaliths at Gobekli Tepe are ornately carved with decorative signs and symbols. Some of them even have highly detailed relief sculptures of animals. So basically animals are carved to look like they are sitting on the pillars, but they are made of the same solid piece of rock. The difficulty of that alone is hard to overstate.

Gobekli Tepe doesn't just represent an engineering and architectural marvel, but an artistic one, as well. Its carvings and sculptures are as expressive as they are detailed. Are we to believe that our hunter-gatherer ancestors emerged from the last Ice Age and that one of the very first things that they built was *this* sophisticated, *this* complex, and *this* beautiful? And this happened just out of nowhere, with no previous knowledge or civilization to draw from?

Once again, I ask—*how?*

Could We Detect An Ancient, Advanced Civilization?

Mainstream archeologists don't have an answer to that question. Though, on the other hand, there admittedly isn't a good answer to their prevailing question which is—*if the builders of Gobekli Tepe weren't just hunter-gatherers and if a more advanced civilization did exist at the end of the last Ice Age, where is the other evidence?*

And that's more than fair.

Yes, we can argue the unique situation created by Gobekli Tepe being intentionally buried 10,000 years ago means that it is almost certainly much better preserved than any potential archeological sites that may have been around at the same time, but without any other analogous sites to point to it's all just speculation.

But is there evidence that it would even be possible for an ancient civilization to just be entirely erased like that?

It turns out that there is. If anything, I'd argue that we tend to grossly underestimate how easily evidence can be lost to the sands of time.

There are many missing chapters of human history—places where the archeological record goes quiet, leaving us to wonder about what transpired. One of the biggest of these gaps occurred before humanity existed, when the ancient hominids from which we are descended still roamed the Earth, between 5 million and 9 million years ago.

That's 4 million years of missing time, the evidence of whatever hominids were doing at the time, how they lived, and how they evolved are essentially erased from the record. Nothing remains.

This is likely because this was a period of Ice Ages. The intermittent cycles of glaciation would have destroyed many potential archaeological sites. And because the sea levels would have been much lower with so much water trapped in ice around the poles, what little record may have remained is likely underwater now.

The paradox of geological time is that 4 million years can be both a very long time and a drop in the bucket. That 4 million years was enough for one branch of the ape family tree to evolve into bipedal, small brained hominids—and yet it's also a short enough period of time that it can be easily lost. It can slip through the cracks and be gone forever.

So if we're talking about thousands of years—tens of thousands of years at most—it's not unthinkable that an advanced civilization could easily rise and fall, leaving little to no evidence behind—especially if they occurred during or immediately before a tumultuous geoclimatic time.

Like say, maybe the event that killed off most species of mammals at the beginning of the Younger Dryas.

And if that were the case, what might still remain of these ancient people? The answer is, probably not much, and what was there would be hard to find—and even harder to conclusively identify. With a little bit of luck, some megalithic structures might still remain, though what state they might be in, we can't be sure.

And there would be stories—tales of a cataclysm deep in our past and of an advanced people whose very existence was swept away by this calamity.

What Happened 12,000 Years Ago?

And then there's the question that we've been dancing around throughout this whole chapter.

You've probably noticed that there is this date that keeps popping up—12,000 years ago, or 10,000 B.C. right at this exact moment in time, give or take a few hundred years, we have three major turning points for the human species—the end of the last Ice Age, the beginning of the agricultural revolution, and the first conclusively dated megalithic structures were built at Gobekli Tepe.

So what the heck happened 12,000 years ago? I submit that there are two possibilities that best fit the evidence.

Human Civilization is much older than we thought.

I find it virtually impossible to believe that a group of hunter-gatherers would have or could have built Gobekli-Tepe without any kind of previous context or reference point. It truly stretched the limits of credulity.

However, the implications of that are startling enough that we may be hesitant to admit it. Because if Gobekli Tepe is, in fact, an artifact of a more advanced civilization with an understanding of engineering, astronomy, art, and perhaps even written language—which evidence at the site strongly suggests—then what we have is this fully formed, complex civilization popping up out of nowhere at the end of the last Ice Age nearly 7,000 years before it was supposed to have existed.

But also, we intuitively know that doesn't really make sense. Civilizations, however they emerge, don't just come out of nowhere. It's a process and a progression, often with fits, setbacks, and false starts that takes literally thousands of years.

The last Ice Age, also referred to as the Younger Dryas, only lasted about 1,200 years—but, as we discussed, this was a rough 1,200 years for anything living in the Northern Hemisphere, with the vast majority of mammals over 40 kgs going extinct within a very short period of time. Whatever caused this massive die off, humans would not have been immune to it, and we almost certainly lost a large percentage of the human population, as well.

So what this implies, is that there could have been some form of advanced civilization that existed on this planet *before* the last Ice Age. On top of whatever brutal conditions lead to the die-off of so many large mammals, rising sea levels would have likely claimed most settlements on the water, which is where early humans tended to build. Perhaps the species sustained heavy damages, their achievements and cities all but erased by the heavy glaciation and the rising sea.

But perhaps those that survived carried this knowledge. Perhaps there were strongholds where life went on, maybe not just like normal, but with at least enough stability to preserve some of the culture of those who came before. And maybe, at the end of the last Ice Age, they built Gobekli Tepe.

And might that explain why after 2000 years of continuous use, they carefully buried it, never to return again? Whatever their reason for leaving, perhaps they still had a memory of the cataclysm and an awareness that even the mightiest of human achievements can be erased from the face of the planet if they are not carefully preserved?

We don't know. We may never know. But based on the evidence, I reject the idea that these are stupid questions. But there is another possibility that is perhaps even more repugnant to mainstream archeologists, which is the idea that:

Humans may have been assisted by an advanced intelligence.

Whatever the origin of the UFO phenomenon, the reality is that, based on the paradigm-shattering technology that these objects display, it is almost certainly an expression of a highly advanced intelligence—whether that be extraterrestrial, ultraterrestrial, interdimensional, temporal, artificial or some other possibility that we haven't even considered yet.

And if the phenomenon is engaging with human beings now, there is no reason to assume that they wouldn't have in the past, as well. In fact, as we discussed in the last chapter, most of these possibilities would almost necessarily mean that they had. And stories of these interactions appear in our art, historical records, and religious texts dating back to the earliest examples that we have, making it impossible to deny the possibility that this is something that humans have been experiencing for thousands of years.

So could it be that the reason that the builders of Gobekli Tepe seem to emerge from the end of the last Ice Age, seemingly out of nowhere, already equipped with the complex set of knowledge and skills necessary to undertake this kind of megalithic building project is because they had a hand? Could they have been gifted this knowledge from a more advanced intelligence?

And if so, what evidence might there be of this profound interaction?

And so that's where we'll pick up in the next chapter, by looking deep into humanity's past to see what evidence exists that might shed some light on these ancient mysteries. And finally, we'll explore the possibility of a major cataclysm during the Younger Dryas and the startling implications it could have not just for our past, but for our future.

10 A Rational Approach To Ancient Aliens [Part 3]: Cataclysms & Non-Human Intervention

When we last left off, we'd established that sites like Gobekli Tepe and the Great Sphinx present major challenges to the current narrative of the emergence of human civilization, pushing back the advent of high civilization by thousands of years. And while mainstream academia has thus far been reluctant to embrace this reality, there is more than sufficient evidence to make the case that human high civilization has its origins sometime *before* the Younger Dryas, more than 12,000 years ago.

This brings to light the astounding possibility that human civilization was in some way seeded, or at the very least guided, by a highly advanced non-human intelligence in our distant past.

The Evidence

But what other evidence is there to support this, admittedly, audacious claim?

Sunken City Of Cuba

In my mind, one of the most flabbergasting archeological discoveries of our times is one that has gotten very little attention, and has been almost entirely dismissed by mainstream academics.

In 2001, Pauline Zalitzki, a marine engineer, and her husband Paul Weinzweig, owners of a Canadian company called Advanced Digital Communications (ADC), were working on a survey mission in partnership with the Cuban government off the eastern tip of Cuba. The area is known to have been the site of several treasure-laden shipwrecks, and ADC was one of a handful of agencies that was working with Fidel Castro's government in an attempt to locate these riches.

The team was using advanced sonar equipment to scan a two square kilometer area of the sea floor when they noticed a series of symmetrical and geometric stone structures resembling an urban complex. There, lurking below the waters, were symmetrical pyramids, the foundations of vast structures, and wide, grid-like boulevards.

In my mind at least, the man-made nature of these structures is undeniable and evocative of the complexes built by the Myans and the Aztecs. But as always, I don't think you should take my word for it. Seriously. Go look at the pictures for yourself right now and decide for yourself.

Once you've seen the pictures, I think you'll see what I mean. It looks like the ruins of a city. And as far as I could tell in my research, no one has so much as ventured a guess as to how these structures could be the result of natural geological processes.

In July 2001, they returned to the site with geologist Manuel Iturralde, senior researcher of Cuba's Natural History Museum, this time equipped with a remotely operated vehicle to examine and film the structures. The cameras revealed even more astonishing evidence including large, perfectly cut stones of granite measuring 8 ft by 10 ft stacked on top of each other.

Despite these findings, Iturralde—who has studied countless underwater formations—was reluctant to draw any conclusions, saying only, "These are extremely peculiar structures, and they have captured our imagination. But if I had to explain this geologically, I would have a hard time."

So why the reluctance to call this submerged city what it so clearly is?

The answer is that, according to the mainstream narrative of human history, it shouldn't be there. Because the last time that particular area was above water was 50,000 years ago—approximately 40,000 years before the end of the last Ice Age and the building of Gobekli Tepe. Which, as you'll recall, is itself a megalithic structure that has been definitively dated to 7000 years before humans were supposed to have been advanced enough to build something of that scale.

At this point, it's glaringly obvious that we don't just have the history of human civilization wrong, but we've undershot its true age by literally thousands, if not *tens* of thousands of years. And yet, mainstream academia simply refuses to engage with this reality, dismissing and discrediting any archeological find that contradicts their dogma.

In the case of the underwater city off the coast of Cuba, the response of mainstream academia has been to rewrite history. After initially refusing to acknowledge the site because its existence was impossible, it seems that some academics have found a workaround. It has now been accepted by some that the ruins are legitimate, but that they are 6,000 years old, not 50,000 years old.

And how did they come to this date?

On the coast near the site is another megalithic structure that was first excavated in 1966 which is believed to be 6000 years old, therefore, the underwater city must also be 6000 years old. However, thus far no explanation has been given as to how the city could have been built on a site that was underwater at the time it was alleged to have been built.

To be clear, I'm not saying that the city is *definitely* 50,000 years old. There is still *a lot* of work to do before anyone could say that conclusively. However, I don't know how anyone could argue that general proximity is a better method of dating than, say, determining the last time that the site wasn't *underwater*. Especially when your entire basis for choosing that younger date is to make it align to your preconceived notions of what must or must not be true.

At any rate, at least this commitment to cognitive dissonance has meant that the site has begun to be explored. And through these explorations they've found structures within the complex as long as 400 meters wide and as much as 40 meters tall. An anthropologist affiliated with the Cuban Academy of Sciences has said that still photos taken from the videotape clearly show "symbols and inscriptions", though the language has not yet been identified.

Puma Punku

Another archeological anomaly that contradicts the mainstream narrative of human history is Puma Punku in Bolivia.

High in the Andes Mountains, in the Altiplano Desert, just southeast of Lake Titicaca, in what is undoubtedly one of the most rugged and desolate places on Earth, lies the ruins of Tiwanaku. Discovered by westerners in 1549 by Pedro Sierra De Leone while looking for the capital of the Incan Empire, Tiwanaku is a massive ancient temple complex.

Within the complex is a large, walled courtyard with carved faces scattered intermittently between the other stones. Interestingly, many of these faces appear to look like humans from different cultures around the world, although supposedly such ancient peoples shouldn't have had any way to be exposed to cultures on the other side of the ocean. And stranger still, some of the faces look humanoid, but still far from human, including one face that looks very much like a gray alien.

And listen, Tiwanaku is cool and all, but what we're really interested in is its mysterious neighbor lying just a half of a mile away—Puma Punku. Puma Punku means "Gateway of the Puma", a name given to it by the local Aymara people who found artifacts depicting warriors wearing masks made out of puma skulls at the site.

So let's start with what's strange about Puma Punku, because the answer is "basically everything."

The site has been destroyed—we're just not sure how.

Puma Punku looks like the ruins of what was once perhaps a large temple complex. What remains are scattered stone blocks and monoliths made from red sandstone and andesite. And these ruins truly are *ruined*. Despite the fact that the largest of these stones weigh up to nearly 150 tons, they are scattered about like a Lego city that just got Gozilla-ed by a toddler.

The site was further corrupted by a misguided and ultimately unsuccessful attempt in 2004 to reassemble the site, so some of the blocks have been haphazardly moved around, making it more difficult to piece together what used to go where.

And it's hard to imagine what kind of force could have strewn these megaliths about like they were children's toys. Due to evidence that some of the stones may have at one time been entirely covered in mud, it's been speculated that perhaps a massive and catastrophic flood could have saturated the ground causing the stone foundations to become unstable before a deluge consumed it, scattering the blocks. But at 12,000 feet above sea level that scenario seems unlikely.

But that's not all that's strange about the ruins of Puma Punku.

We don't know how the builders of Puma Punku got the stones to this site.

Based on detailed petrographic and chemical analyses of samples from individual stones and known quarry sites, archaeologists have determined that the red sandstone blocks at the site were transported up a steep incline from a quarry near Lake Titicaca roughly 10 km (or 6.2 miles) away. And if that wasn't daunting enough, the andesite blocks were determined to have come from quarries within the Copacabana Peninsula about 90 km (or 56 miles) away.

Archeologists still don't have a definite answer on how this was done.

However, I will note that I'm always a little less swayed by the arguments that megalithic blocks were too big to be moved great distances. It's interesting, for sure. And I would say that the size and the distance of the blocks being moved could speak to the advancement and wealth of that culture because the resources needed to undertake such a thing would be incredible.

However, there are two reasons that I hesitate to assume an advanced technological answer to explain megaliths being moved over large distances:

The first is that there have been various studies and experiments that have shown that there are some pretty ingenious ways of moving these megaliths, albeit very slowly, including using log rollers and other basic materials that would have been available—and we can't rule out that ancient people were able to figure it out. And the second is that I think we tend to underestimate what can be accomplished by a civilization that is willing to throw endless amounts of human suffering at a problem.

So while it's certainly an interesting data point, I'm not convinced that it can be used as decisive proof of the use of advanced technology or the possession of special knowledge.

The stonework at Puma Punku shows evidence of having been done by advanced tools.

The stonework at Puma Punku is bananas. I seriously don't even know how else to say it.

Red sandstone and andesite are both very hard kinds of rock. To put that into context, we use a scale called the mohs scale to measure the hardness of different minerals on a scale from 1-to-10, with 1 being the softest and 10 being the hardest. Diamonds are a 10 on the mohs scale. Granite is a 6. By comparison, red sandstone is a 6 or 7 and andesite is a 7.

Given the hardness of these stones, the stone work at Puma Punk defies any traditional explanation. The stones were cut in such a precise way that they fit perfectly together, locking into place without the use of mortar. Not even a razor blade can slide between the rocks. To accomplish this would not just require astounding technical precision and finesse, but an understanding of descriptive geometry.

The architectural historians Jean-Pierre and Stella Nair who conducted the first professional field study on the stones of Puma Punku concluded about these stones that:

"[...] to obtain the smooth finishes, the perfectly planar faces and exact interior and exterior right angles on the finely dressed stones, they resorted to techniques unknown to the Incas and to us at this time. The sharp and precise 90° interior angles observed on various decorative motifs most likely were not made with hammerstones. No matter how fine the hammerstone's point, it could never produce the crisp right interior angles seen on Tiahuanaco stonework. Comparable cuts in Inca masonry all have rounded interior angles typical of the pounding technique [...]. The construction tools of the Tiahuanacans, with perhaps the possible exception of hammerstones, remain essentially unknown and have yet to be discovered."

And those tools remain undiscovered until this day.

Even more astounding is the fact that many of these blocks are finished to 'machine' quality and with tiny holes drilled to perfection. And these holes don't look like decoration, but like utilitarian means of construction and joining these pieces together. The machined appearance of these stones is all the more compelling because to create stonework with a similar level of precision out of the same materials today, we'd need to use advanced equipment like lasers and diamond-edged circular saws.

Further adding to the mystery is that many of these stones aren't just precisely cut into complex, multi-faceted blocks, but they are done *so* precisely as to be completely identical, allowing one block to be perfectly interchanged with another, almost like the pre-fabricated building materials we use today.

And interchangeability of parts, as seen in the blocks of Puma Punku, is something that on its own is suggestive of a more advanced society. While there are a few isolated examples of interchangeable parts being used in the ancient past (almost always in weapons to make warring easier—go figure), for the most part, interchangeable parts weren't something that was commonly used before the early 1800s.

Many of the stones are magnetized.

Apparently if you hold a compass to the stones the needle will spin and spin. Why would this be? What would cause this? Could it have been some result of the construction process, or the result of a natural phenomenon? We simply don't know.

And what's extra frustrating is that we will likely never know how the builders of Puma Punku did it, because we don't even know who they are for sure.

No one knows who built it.

While Puma Punku was an important site for the Incan people, they didn't build it. Mainstream archeologists have offered several different potential dates for the creation of the site, with the most commonly accepted date being around 500 A.D., which predates the Incan empire by more than half a millenia. When the earlier Spanish conquistadors asked the Incan ruler who built Puma Punku they claimed that it had been built by the gods in only one night.

We don't know how old it is—but there is evidence that it's much older than believed.

And while mainstream archeologists insist that the site is no more than 1500 or so years old, there are many who question this date. Not the least of which is the local Aymara people, whose local elders and historians claim that the site is 10 times older than that, having been built 15,000 years ago—or 3000 years before the end of the last ice age.

Austrian archaeologist Arthur Posnansky who was one of the first archeologists to study Puma Punku and who spent decades at the site also advocated for the date of 15,000 years ago citing various astrological alignments that would have tied it to that time.

And while I find the evidence that Puma Punku pre-dates the ice age to be compelling, it doesn't need to be true for Puma Punku to be unexplainable within the current narrative. Whether it was 1500 years ago or 15,000 years ago, humans simply didn't have the tools or the knowledge to create megaliths of such complexity and precision.

So *who* built it and *how* are the questions that really matter. The *when* is just an interesting detail.

Baalbek

On the other side of the world, in Baalbek, Lebanon lies another archeological site that defies explanation.

We don't know for sure who ordered the construction of the Temple of Jupiter at Baalbek, nor do we know exactly when it was built. However, most archaeologists date it to either the first century B.C. or the first century A.D. The style of the temple with its towering columns and its dedication to the Roman god Jupiter, make it undoubtedly a Roman construction, hazy though its origins may be.

However, what's interesting about Baalbek isn't so much the Temple of Jupiter, but the massive stone platform on which it sits. And although mainstream archeologists date this platform to the first century B.C., more or less aligning with the building of the Temple of Jupiter, I'd argue that it's pretty obvious that is not accurate.

First of all, the foundation of the temple itself appears to have been built on the foundation of an older temple. And both weathering and logic would suggest that the massive stone platform that it sits upon must be even older than that.

But once again, whether you date this structure to the 1st century B.C. or the 10th century B.C., it doesn't matter—because at either time humans should not have had the ability to build it.

Now, I've already mentioned that I tend to not find the "rock too big, can't lift" argument to be very compelling. But the megaliths at Baalbek are a clear exception. Integrated into the back wall of this massive platform are three of the biggest megaliths known to humankind. Each of these stones weighs at least 800 tons. To put that into perspective, that is the same weight as *five* average adult blue whales. It's insane.

And not only were these monstrous megaliths quarried and moved to the site, but they were then somehow lifted 20 feet in the air to rest in their current positions. All you really need to do is look at a picture of these rocks to understand that something we don't understand happened at Baalbek.

But those megaliths aren't even the largest found in the area. Three megaliths, conventionally known as the "Stone of the Pregnant Woman" , the "Stone of the South", and the "Forgotten Stone", lie nearby and are the largest stones ever to have been quarried. At a staggering 1000, 1200, and 1600 tons respectively, the size of these megaliths boggles the mind.

Now granted, the location of these three massive stones gives every indication that they were simply too big. All 3 of them still lie in the quarry where they were cut, and there they shall likely remain for millions of years. Even now our ability to move them would be extremely limited and would require an insane amount of time and resources.

The Most Impressive Parts Of Megalithic Structures Are Usually The Oldest Parts

And what we see at Baalbek is a perfect example of something very surprising that we see a lot with megalithic structures, which is that the oldest parts of the construction are by far the most advanced.

We see it at Gobekli Tepe. We see it at Tiwanaku where Puma Punku is located. And there are particularly stunning examples of this in Peru.

The ruins of Coricancha—which means, *The Golden Temple*—can be found in Cusco, Peru. In many places, all that remains of this ancient Incan holy site is its massive megalithic foundation, upon which a Spanish church was built in the 16th century.

However, as beautiful as this 16th century church is, it looks somehow sad and small on top of the foundation of Coricancha. The precision with which the megaliths are cut and fitted together, including a perfectly curved retaining wall is enough to take your breath away. Once again, I'd recommend that you look at the pictures to get the full impact.

We can see something similar at the Peruvian site of Machu Picchu. The oldest and most foundational parts of the site, once again, show the most advanced knowledge and the largest megaliths, with subsequent rounds of building being done on top with much smaller stones and far less precision. This has caused some to suggest that Machu Picchu may not have been originally built by the Incan people, but rather repurposed and rebuilt by them on a far more ancient site.

And whether or not that is the case, it defies logic that the Incan people would start out with such a high-level ability to create megalithic structures of this scale and precision, only to be rebuilding the walls a few hundred years later with much smaller blocks of stone.

Much like Gobekli Tepe, we're again and again confronted with an odd situation. An advanced civilization of megalithic builders pops out of obscurity with little to no evidence of ramp up time, does their most breathtaking and awe-inspiring work, before the civilization begins a long, slow decline, each new wave of building being only a shadow of what came before.

This doesn't make sense. We know that it doesn't make sense.

What Does It Mean?

So what does this all mean? What conclusions are we to draw from these anomalous archeological sites that can be found around the world?

The examples that we've explored are just a handful of examples, but they are representative of the four main problems that these sites present.

The first, is that many of them—like Gobekli Tepe in Turkey, the Sphinx in Egypt, and the sunken underwater city off the coast of Cuba — are simply way too old. They show a level of sophistication—not just in construction and engineering, but in art, geometry, and astronomy—that are thousands, if not *tens* of thousands, of years ahead of their time.

The second, is that some sites, like Puma Punku in Bolivia, show clear evidence of machining and perhaps even prefabrication. We simply have no way to explain how ancient peoples could have achieved with primitive tools what it would take advanced, laser guided and diamond tipped tools to accomplish now.

The third, is that some of these sites have megaliths that are far too big to have plausibly been moved without some form of advanced technology. As I've said, I generally don't find arguments about megaliths being too large for ancient peoples to have quarried and moved to be particularly compelling. If you don't care how long it takes and you don't care how many people die, humans can do some pretty unbelievable things.

However, there are exceptions. The enormous megaliths of Baalbek are one example, but there are others of nearly equally gargantuan proportions in Egypt and Peru. There's even a site in Southern Siberia called Gornaya Shoria that—if it can be proven that they are, in fact, man made and not a naturally occurring geological structure—has megaliths that outweigh even the monsters found at Baalbek. I'm less convinced on that one, but the photos of the site are undeniably compelling.

And finally, we have the mystery that we see all over the world— from Gobekli Tepe to Egypt and from Baalbek to Machu Picchu—where the earliest megalithic structures that a culture creates seems to be, by far, its most advanced and sophisticated with subsequent builds and restorations never coming close to rivaling that of the first builders. Again and again with megaliths we see advanced high culture seemingly appearing out of nowhere at the very height of its power and sophistication with little evidence of a ramp up period.

So who were they?

Answering that question only gets more complicated when we layer on two more puzzling observations:

These pockets of civilization were both relatively isolated and anomalous *and* spread all over the planet.

Regardless of what any of the archeological evidence surrounding these megalithic sites may suggest, there are still certain facts that we can't deny. And the most glaring of these is that prior to 6000 years ago—and certainly prior to the last Ice Age over 12,000 years ago—most humans were still hunter gatherers. We have found their camps, their tools, their clothes, and their art everywhere on the planet where evidence of homo sapiens can be found.

And this only makes these megalithic sites that hint at advanced civilization in our distant past more mysterious. It raises the startling possibility that while most of humanity was still living a lifestyle that was akin to that of our early hominin ancestors, there existed pockets of civilization where humans had attained a dazzling level of understanding of engineering, geometry, astronomy, and more that rivaled—and in some cases may have exceeded—our modern understanding.

In some ways, this is almost the inverse of what we see on Earth today. Today most of the world is at least somewhat modernized. 86% of people worldwide can read and write, and an almost identical proportion have access to electricity. However, there are estimated to be between 100-200 uncontacted tribes on the planet who have had little to no interaction with the modern world and who are still essentially living the way that our ancestors did hundreds and even thousands of years ago.

And while it can initially be shocking to realize that there can be so much variability in how humans living at the same time might be living their lives, the situation in which we find ourselves in our modern world at least makes some level of intuitive sense. We understand how it could be that our modern lifestyle hasn't reached people living in remote places. And when we look at the history of colonialism, it makes sense that there are tribes in places like the Amazon who may have gone far out of their way to avoid interacting with outsiders with strange new weapons and diseases.

But looking into our distant past, it's frankly confounding that such a sophisticated level of high civilization could be achieved by a few while the vast majority of humans were still living as hunter-gatherers. What accounts for this difference? How did they get so far ahead? What happened to them that was so fundamentally different than what happened to everyone else to account for such a massive disparity?

And this mystery only deepens further when we recognize that this level of advanced civilization didn't just appear in one corner or region of the world—but in pockets scattered across nearly every continent. And, for reasons that we still don't understand, in each of these far flung cultures the attainment of this knowledge seems to coincide almost exactly with the building of megaliths.

And as if things weren't confusing enough, we have one added layer of nuance and complication which is that:

These civilizations show evidence of advanced technology that does not fit within our known paradigm.

As we discussed previously in part one of this series, if humans were to entirely disappear tomorrow there would be surprisingly little left to tell the tale of our modern lives. Our cities and skyscrapers, while they look impressive, are not built to last. In fact, it would take less than a thousand years after the disappearance of humans for all of New York City to become a pristine forest again.

And because we've only been at this relative level of advancement for around 100 years—barely a blip in terms of the dizzying scope of geological time—before long, it would be almost impossible to find anything of our way of life in the archeological record. It's surprisingly easy for an entire civilization, even one as advanced and global as our own, to disappear without a trace.

But there is one form of evidence that we will leave behind that will leave an indelible mark on the planet for hundreds of millions of years, and that is our use of plastics.

Plastic never biodegrades, which means that every single piece of plastic that has ever been created—unless it got shot up into space at some point—is still here on the planet. And it will be. Forever. The Empire state building may vanish without a trace in a matter of centuries, but your toothbrush could be discovered virtually intact by someone millions of years in the future.

We use around 100 million tons of plastic every year, at least 10% of which ends up in the ocean. This means that our current time period will be marked forever in the geologic record with a layer that is basically speckled with plastic, even in the very deepest parts of our oceans.

It's already happening. In Hawaii they recently identified a new kind of rock that has been washing up the shores called a plastiglomerate, which is essentially rock that has become fused with melted plastic.

So if these civilizations were in many ways as advanced as we are, and if they perhaps persisted for thousands of years at a time, why don't we see something similar in the geologic record?

It's certainly difficult to imagine our modern lives without plastics. Beyond just convenience products, plastics play a critical role in healthcare, safe food storage and packaging, complex scientific equipment, the space program, and more.

But what's clear is that, whatever level of advancement may have been achieved by our ancestors in the distant past, it apparently lacked many of the trappings of our modern lives. And if their abilities and tools did truly reach the level of what we would call "technology", as is suggested by megalithic sites like Puma Punku and others, it was likely of a fundamentally different kind than what we have today.

These civilizations didn't use plastics, and they almost certainly didn't use fossil fuels. This begs the question of how their technology differed from our own, and it offers us a tantalizing hint at a potential technological paradigm that may have once existed that was perhaps in better harmony with the environment than the one we currently occupy.

The Possibilities

As we can see, each piece of this puzzle is baffling in its own right, and when we put it all together the path to the truth becomes even more labyrinthian. So what could explain all of this bizarre evidence of advanced civilization in our distant past?

If we're willing to put aside the established narrative of human civilization as told by mainstream academia and consider the evidence with an open mind, a surprising narrative begins to emerge.

Human Civilization is far more ancient than we thought.

At this point, I don't think that's even up for debate anymore, though mainstream academia may not yet be ready to accept it. As we discussed in

The UFO Rabbit Hole

the last chapter, Gobekli Tepe alone presents enough evidence to prove that we've underestimated the dawn of human civilization by, at minimum, 7000 years. And the sunken city in Cuba hints that we may still be off by tens of thousands years more.

This revelation is astonishing on its own, but it unfortunately creates more questions than answers. The reality is that the further we push back the start of human civilization, the more unlikely it becomes that any group of people could have risen to that level of sophistication on their own. It just doesn't make sense.

After all, as we discussed in the last chapter, the oldest sections of Gobekli Tepe were built approximately 12,000 years ago, near the end of the last Ice Age, also known as the Younger Dryas. And, as with many other megalithic structures, the oldest sections are the most sophisticated. Gobekli Tepe also clearly shows evidence that its builders had a complex understanding of engineering, mathematics, geometry, and astronomy. That kind of knowledge doesn't just come out of nowhere and it should have taken thousands of years to achieve that level of mastery.

The Younger Dryas, which lasted around 1300 years, was a horribly tumultuous time for the planet with over 82% of mammals over 40 kg dying off in North America and suffering similarly massive losses elsewhere around the globe. It seems unlikely that humans could have made those sorts of advancements during such a chaotic and dangerous time period—which strongly suggests that the origins of high civilization lie somewhere *before* the Younger Dryas.

But how could this possibly be?

This suggests that while the vast majority of our early human ancestors were still living in caves and hunting mammoths with spears, that someone, somewhere was inventing geometry and astronomy and written language. And honestly, I can understand why mainstream academia has been so resistant to embracing this reality, because it simply doesn't make sense. It's not possible.

Unless...

Unless we are willing to admit the possibility that we may have been gifted this knowledge and that early human civilization may have been intentionally seeded.

But by whom?

I'm not saying it was aliens, but...

It could have been aliens.

And, as I'm sure you've already guessed, when I say "aliens" I'm speaking broadly about a non-human, intelligent species that could have a number of potential origins whether it be extraterrestrial, ultraterrestrial, interdimensional, or extratemporal. Where they came from and who they were is a matter of debate, but there seems to be no way around the fact that at some point in the distant past humans were helped by a very advanced and intelligent *someone*.

It would explain the sudden emergence of mathematics, science, and high civilization long before such things ought to have been possible.

It would explain how these advanced megalithic civilizations seem to continually spring out of nothingness and into the height of their power only to slowly degrade over time, each new iteration a copy of a copy of a copy until its original glory is lost.

It would explain the seemingly impossible megalithic sites like Puma Punku where the stones are cut too perfectly, and Baalbek where the stones are far too big.

It could even explain an advanced technological paradigm that is fundamentally different from the one in which we currently live, and which seemed to exist without the vast networks of infrastructure and resources that are required to support technology in our modern world.

And, perhaps most tellingly, it would explain the eerily similar myths, legends, and religious texts from around the world—dating back to the oldest records that we have—that tell the tale of powerful beings who came down from the sky and gifted humans with wisdom and built mighty civilizations that they ruled over for thousands of years.

And might it also explain that startling claim made by Tom Delonge about the idea that he believed was responsible for not just gaining his advisors attention, but gaining entrance into their shadowy world?

If you'll recall, Delonge said multiple times in those early and unguarded 2016-2017 interviews that while it was his pitch that got him in the door with his advisors, it was something else that made them take him seriously—and that was the idea that the UFO phenomenon represents multiple "gods with a little *g*" who war with each other, and who have, for millennia, controlled and influenced humanity.

Could all of this be true? And if, as Tom Delonge has said, there are those within our government who know about this, take it very seriously, and who recognize that it is related to the UFO phenomenon, what does that mean about—you know...everything? Reality? Like seriously, what the fuck?

And, if I'm being honest, this is where I start to feel like I'm losing it. Because I keep retracing my steps, again and again, trying to figure out where I went wrong to end up with these same conclusions that seem to be somehow both entirely obvious and entirely impossible.

And I'm putting all of this out there knowing that I could be wrong. If you pull the wrong brick out of this game of "Reality Jenga", it would all come crashing down. I accept that, and I'm genuinely at peace with it.

But I will say this:

In many ways the path we're taking down the rabbit hole mirrors the path that I took. And there's a reason that we've ended up spending so much time on this topic, and why I'm only now ready to say that there is significant evidence of intervention by an advanced non-human intelligence in humanity's distant past.

Because this is where I started, and when I first pursued this line of questioning, I fully expected to debunk it—and easily.

If you go looking for literature and research on UFOs, you'll immediately find that there isn't much to find. And sorting through and weighing the ultimate credibility of what *is* there is daunting, if not downright impossible. This topic is shrouded in secrecy and insulated by stigma to the point that there is no bedrock of consensus truth, there is no clearly unimpeachable source material, there's not so much as a sticker saying "this end up" slapped on the side of ufology.

But history—now that was something we had a handle on. And if there truly was a non-human intelligence of some kind that was interfering in human affairs, surely there would be evidence. And I'm not just talking about myths and legends. I'm not even talking about megalithic stones and whether or not they were too big to lift. All of that can be refuted and none of that gives enough of a foundation of certainty from which we can begin to build.

What I'm talking about is clear evidence that the path of human history has been in some way diverted or guided in ways we can't explain by something bigger than ourselves. And no one was more surprised than me when I found exactly that. And I'd argue that the evidence is overwhelming.

Who Is Inside The UFOs?

If we accept that human civilization emerged as the result of intervention by a non-human intelligence in the distant past, and that this event is related in some significant way to the UFO phenomenon, this creates an interesting question about the UFO phenomenon as we experience it today, which is— who is inside the UFOs?

Well the most obvious answer would be that it's the Others— whoever they are. Perhaps they are some version of the gods of old: more-or-less immortal beings with technology so advanced that—even to our modern eyes—it looks more like magic than science.

And, in chapters 3, 4, 5, we talked through several different possibilities for who exactly they might be and where they might have come from.

But with the revelation that human civilization could be thousands, if not *tens* of thousands, years older than we previously thought, and with so much evidence pointing to the startling fact that some of our ancient ancestors may have at least been exposed to advanced technology which defies our current understanding, it suggests a profound new possibility:

What if some of the UFOs that are being reported in our skies and moving through our oceans belong to humans? Could what we're seeing be explained by some isolated offshoot of humanity that somehow kept the flame of civilization alive across the millenia, and that in so doing, got a technological head start on us of ten thousand years or more?

It's a fascinating possibility.

Why Did Ancient Humans Create Megaliths?

But there's one mystery, more than any other, that has turned me inside out and kept me up, late into the night until dawn streaks the ceiling. In my mind, the question that matters isn't *who* built the megaliths or even *how* they accomplished it. The only question that matters is *why?*

Why would people who were only recently hunter-gathers, or who perhaps *still were* hunter-gatherers, do this?

The scale of these sites and the megaliths that our ancestors cut to raise them is enough to take our breath away—even now. Their symmetry and perfection make us feel small. In their presence we speak in hushed tones.

How did people who spent all of their lives in the natural world, a place that—while full of grandeur and unspeakable beauty—is utterly devoid of the right angles, the precision, and the elegant geometry made manifest in these megalithic temples?

And we assume them to be temples or sacred religious sites even when we have no other evidence for calling them such because in them we see what can only be called *worship*. Megaliths are worshipful. We know that in our bones.

What did our ancestors see? What did they believe to be true? What made them strive to defy the very limits of not just human ability, but of our capacity to recognize and create both divine order and transcendent beauty?

And where did that capacity come from? When did it arise and how? And what sparked that ember of knowledge? And what fanned that flame to cause the fever of megalithic building that seemed to consume our ancestors in every far-flung corner of the globe?

Did the stars call to them? Did gods walk among them?

Or perhaps they raised the megaliths because they knew something that those who build skyscrapers have forgotten—that everything is temporary, that the full sweep of human history is but a mote floating in the infinite cosmic wind, that mighty civilizations can rise and fall leaving little more than a whisper of their greatness—and that, in the end, only the mighty megaliths remain.

If They Existed, What Happened To Them?

Hidden at the center of this story about the dawn of human civilization is another near-forgotten chapter of our history—a devastating global cataclysm in our distant past that nearly extinguished the light of human civilization forever.

But what happened 12,000 years ago? And could it happen again?

There are a few different theories, none of which is perfect, but all of which are super interesting. So what are they?

The Younger Dryas And The Ice Ages

As we've discussed, the Younger Dryas refers to the last mini-ice age that occurred roughly between 12,900 and 11,600 years ago. There are lots of common misconceptions about ice ages, from what they are to how long they last, so it's worth taking a minute to understand exactly what we mean when we say "ice age".

We tend to think of an ice age as one event that lasts for a few thousand years and then ends, but an ice age is actually part of a much longer cycle.

An ice age is a long interval of time, usually spanning millions to tens of millions of years, when global temperatures are relatively cold and large areas of the Earth are covered by continental ice sheets and alpine glaciers. However, within an ice age are multiple shorter-term periods of warmer temperatures when glaciers retreat (called interglacial cycles) and periods of colder temperatures when glaciers advance (called glacial cycles).

At least five major ice ages have occurred throughout Earth's history. The first was over 2 billion years ago. The most recent one began approximately 3 million years ago, and technically, it continues to this day. What we are experiencing now is one of the interglacial cycles within the larger cycle of the ice age.

This interglacial period began about 11,000 years ago with the end of the Younger Dryas. Before that, the last period of glaciation, which is often informally called the "Ice Age," peaked about 20,000 years ago. As for when the next one will be—we're not actually sure.

We know that ice ages are triggered by a variety of complex geologic and environmental factors including changes in oceanic and atmospheric circulation patterns, varying concentrations of carbon dioxide in the atmosphere, and even volcanic eruptions. We suspect that the current ice age began when the land bridge between North and South America (AKA the Isthmus of Panama) formed, ending the exchange of tropical water between the Atlantic and Pacific Oceans, which significantly altered ocean currents.

But, with climate change escalating at an increasing and unprecedented rate, it's hard to tell what impact this may have on the coming glacial cycles, making it difficult to impossible for scientists to determine when the next one might occur.

What Caused The Younger Dryas?

However, none of this really helps us understand the Younger Dryas. Whatever caused the Younger Dryas, it wasn't as simple as changing ocean currents, and the devastation that ensued is evidence of something far more profound.

As you'll recall from the previous chapter, at the onset of the Younger Dryas there was a massive, worldwide extinction of large mammals weighing over 40 kg (or 88 pounds). It is estimated that 82% of these animals disappeared in North America, 74% in South America, 71% in Australasia, 59% in Europe, 52% in Asia, and 16% in Sub-Saharan Africa. This mass extinction event marked the demise of the mammoths, as well as the disappearance of horses in North America, and other species, including bison, deer, and moose, suffered massive population losses.

This kind of near-global extinction event is not consistent with the simple onset of an ice age, which tends to occur gradually—allowing animals to migrate and generally limiting population loss to species and areas within the glacial regions. And yet, this is the story that is still embraced by most mainstream academics—that the Younger Dryas was caused by changes in ocean currents.

So what happened?

One of the most popular and well-supported theories is that there was some sort of massive cataclysm, a black swan event of some kind that rapidly and drastically changed the conditions on the planet leading to a massive die-off of terrestrial mammal—but there is considerable disagreement about what this might have been.

So let's talk about some of the most popular theories.

A Comet Impact

Many studies have shown significant evidence of a major biomass burning event that occurred right at the beginning of the Younger Dryas. So basically, something happened that caused so much of the planet to catch on fire that we can literally see it in the geologic record.

At sites around the world we've found a thick black layer, often called the Younger Dryas Boundary, that is consistent with biomass burning right at the depth that would align with the beginning of the Younger Dryas, emphasizing how wide-spread this disaster truly was.

In ice-cores from Greenland, Antarctica, and Russia we can see drastically increased dust concentrations, as well as chemical signatures consistent with massive biomass burning. Whatever happened, it left its mark across the globe.

For many, the best explanation for this would be a comet impact—whether it be one comet or many pieces of a comet that may have broken apart—that struck somewhere in the northern hemisphere, most likely in North America.

One study published in the *Journal of Geology* estimated that as much as 9% of the Earth's terrestrial biomass burned in one devastating event. But after the fires died, things would have cooled down quickly. The amount of dust and ash in the atmosphere would have blocked out the sun causing temperatures to plummet, triggering a new ice age.

A Solar Event

We tend to think of our sun as being constant and stable, but it's actually anything but. It goes through periods of solar maximums, where it is very active and erratic, resulting in more activity like sunspots and solar flares, about every 11 years. We're generally protected from this sort of activity by the electromagnetic field that surrounds the Earth, but like our sun, this field also goes through cycles where it is sometimes stronger and sometimes weaker.

Without getting too in the weeds, it could be that a particularly violent solar maximum could have coincided with a weakening of the electromagnetic field around the Earth which could have caused massive electrical storms and allowed deadly cosmic rays and radiation to reach the Earth. This could also account for much of the same evidence as a comet strike.

And to be honest, I don't think that I have a firm enough grasp on the complex geo-climatological factors involved to give an opinion one way or another. Suffice it to say, the evidence that the Younger Dryas was initiated by a devastating burning event is overwhelming, but what caused it is still a matter of some debate to be settled by people much smarter than myself.

The Mystery Of The Poles

However, there is another set of much more bizarre theories about not just the cataclysm that occurred at the onset of the Younger Dryas, but a long cycle of cataclysms that have occurred on this planet—and bizarre though they may be, I promise if you hear me out I'll make the case for why this is worth your consideration.

So let's look at the evidence.

The Flood

When we look at the earliest stories and myths in cultures around the world, we find one story that appears again and again—which is that of a great flood. Much like Noah's flood in the Bible, these stories tend to follow a common course. There is a great flood that is sent by one or more gods to destroy the Earth. Only a handful of people are saved, usually by some kind of warning or divine intervention, and after the flood the few survivors repopulate the Earth.

And because of the ubiquity of this story, there are many who argue that it must have been a flood that was to blame for the near erasure of whatever human civilization existed before the Younger Dryas.

And I will say that through this process one thing that I truly didn't expect, but that has been impressed upon me by the evidence again and again, is that I don't think we give nearly enough attention to ancient myths and legends when it comes to understanding our distant past. I'm not suggesting that we should treat these as historical records, but we shouldn't dismiss what they have to say out-of-hand either.

Stories that were important enough to pass down through the ages, to record in the pages of our religious texts, and to literally chisel into the stone walls of megalithic temples were important to people for a reason. They may or may not convey literal truth, but they do reveal something about the people telling the story. And at the very least, I don't think it's crazy to argue that the fact that this story of a massive flood that nearly wiped out humanity shows up in cultures around the world following a nearly identical narrative arc suggests some sort of shared experience or awareness on the part of our ancestors.

Let's stop for a minute and just say for the sake of argument that a massive, global flood did happen in the distant past. Wouldn't we expect to see exactly what we're seeing? Wouldn't we expect it to be recorded in myth and legend across cultures?

Now, of course, that doesn't prove anything. *And* it doesn't really help us to overcome the most obvious and logical refutation of the reality of the Great Flood—which is that, as far as we know, a global flood on a scale that could nearly wipe out humanity doesn't seem like it should be possible.

In the myths, the flood was usually caused by one or more deities. But if we're looking for an actual known, non-supernatural mechanism by which a flood of this scale could have happened, we come up short.

We're familiar with how floods happen, but they tend to be regional, and wouldn't explain the global impact that the stories point to.

We're even familiar with rising sea levels and how entire cities in low-lying coastal areas can end up underwater due to climate change. But, looking at our modern day situation, even the most catastrophic estimates for sea level rise by the year 2100 are at around 8 feet. Which, don't get me wrong, would be absolutely devastating—but it wouldn't happen overnight. It wouldn't be the great deluge described by our ancestors.

So if the Great Flood was real, is there anything that could possibly explain it?

There just might be.

Albert Einstein & Geological Cataclysms

And one theory describing this bizarre cataclysm found an unlikely proponent in one of the great minds of his or any time—Albert Einstein.

Due to the scope and impact of his work in the world of physics, most people don't realize that Albert Einstein also had a considerable interest in the earth sciences, and made important contributions in that field, as well.

In particular, he did some important work in the area of fluvial geomorphology—which is the study of landforms and processes associated with rivers. Einstein was the first to articulate how helical flow helps determine meander length and promote down-current migration of the meandering rivers. So basically, it was a model for how rivers form and cut their path across a landscape.

Einstein was also very interested in emerging theories of the time including Charles Hapgood's ideas about crust displacement.

What Hapgood proposed was that the uneven distribution of weight caused by accumulation of ice around the Earth's poles could, at times, cause the Earth's crust to slip dramatically on its mantle by as much as 30 degrees. The impact of such an event would be absolutely catastrophic and would result in everything from volcanic eruptions to massive flash floods worldwide. It would also cause the poles to change positions as the land that used to be over the poles would be shifted thousands of miles away.

Hapgood presented extensive climatological research that he claimed—and which Einstein agreed—supported his hypothesis that global patterns of climate change over the past 100,000 years could be the result of displacements of the Earth's crust and corresponding shifts of the geographic poles. In modeling the data he proposed that three geographic pole shifts had occurred over the past 100,000 years.

According to Hapgood, approximately 100,000 years ago, the north pole was in the Yukon. Then around 75,000 years ago, it shifted to a spot between Iceland and Norway. Around 50,000 years ago it shifted from there to the Hudson Bay. And then finally, sometime around 12,000 years ago, it shifted to its current location.

And there's that date of 12,000 years ago again.

Now, the mainstream scientific stance is that this is impossible, and that the shifting of the poles does happen, but very slowly — approximately 30 degrees over the past 200 million years. And I honestly hope that they're right on this one, because I frankly find this idea to be terrifying. Einstein, however, was super into it.

Over the last couple of years of his life, Einstein maintained an ongoing correspondence with Hapgood about his ideas, and was very encouraging of his efforts. And while he did eventually convince Hapgood that the weight of polar ice wouldn't be sufficient to shift the Earth's crust, he was compelled by the evidence that the poles had—*somehow*—shifted dramatically.

Einstein was such a fan of Hapgood's work that he actually wrote the foreword to his book, *Earth's Shifting Crust*, and when Einstein passed away it was that very same book that still sat open on his desk.

But it wasn't just Hapgood whose work caught Einstein's attention. An even more bombastic theory about cyclical global cataclysms was put forth by a man named Immanuel Velikovsky in 1940 in his book, *Worlds In Collision*.

Velikovsky compared numerous natural disasters in the Bible and other ancient texts and based on that he proposed the following truly wild scenario:

According to Velikovsky, at some point in the past Jupiter *somehow* burped out a comet that then fell toward the sun in a long elliptical orbit. As it went past Mars it pulled it out of its orbit and also yanked off its atmosphere. It then passed the Earth causing a series of catastrophes, which repeated approximately every 52 years as it made its long orbit around the sun before eventually settling into an orbit between Mercury and the Earth, becoming the planet we know today as Venus.

Now, I'm sure I don't need to tell you that there is literally nothing in all of science that would support that. The entire thing is absurd and impossible for a multitude of reasons that a fifth grader could likely point out. So why was Albert Einstein wasting his time on these crazy ideas?

Well, from what we find in the record of correspondence between the two men, Einstein believed that there was at least part of his idea that had merit. He wrote the following in a letter to Velikovsky about his book:

I have read the whole book about the planet Venus. There is much interest in the book which proves that in fact catastrophes have taken place which must be attributed to extraterrestrial causes. However, it is evident to every sensible physicist that these catastrophes can have nothing to do with the planet Venus...Your arguments in this regard are so weak as opposed to the mechanical–astronomical ones, that no expert will be able to take them seriously. It were best in my opinion if you would in this way revise your books, which contain truly valuable material. If you cannot decide on this, then what is valuable in your deliberations will become ineffective, and it may be difficult finding a sensible publisher who would take the risk of such a heavy fiasco upon himself.

Or, as he said more succinctly in a later letter regarding Velikovsky's work:

I can say in short: catastrophes yes, Venus no.

Einstein's rather direct feedback didn't seem to have a negative impact on his relationship with Velikovsky, and Einstein continued to encourage him in his work throughout their 9-year correspondence.

So what becomes clear is that, although Einstein wasn't compelled by their explanations for the exact mechanics of *how* these cataclysms occurred, he was convinced by the work of both men of the reality of cyclical global cataclysms in our distant past.

Now granted, just because he was one of the greatest thinkers in human history doesn't mean that Einstein was infallible. And just because he took an interest in the topic doesn't mean that it necessarily is true. But it is interesting.

And Einstein isn't the only notable to have taken an interest in this idea—the C.I.A. has actually shown an interest, as well.

The CIA and *The Adam & Eve Story*

In 2013, among a massive dump of newly declassified CIA files was something very strange—a "sanitized" version of a book called *The Story of Adam & Eve*.

This book was written in 1966 by a former U.S. Air Force employee named Chan Thomas. But before the book was released, it was seized, banned, and classified by the C.I.A.

So what could this book possibly say that would make the CIA take such an interest—*and* go to such great lengths to keep the public from reading it? I've read the declassified version, and I've gotta tell you—it's more than a little perplexing.

The book describes a series of disasters that have occurred on the planet at intervals of approximately every 7000 years. These disasters are caused by the Earth's crust becoming unbalanced and slipping along the mantle, much like in the crust displacement theory, but this is on another level. Instead of the crust slipping 30 degrees or so, *The Story of Adam & Eve* describes a scenario in which the poles would swing basically all the way down the equator in a matter of hours.

According to the book, when this happens, the oceans and atmosphere continue to spin in the direction they always have while the crust moves independently. This causes catastrophic 1000 mile per hour winds, and even more disturbingly, the oceans continue their journey west-to-east with the spinning of the Earth, while the crust is jerked north-to-south, which—according to the book—would cause a massive, 2-mile-high wave.

In the U.S., this wave would start on the west coast and move east across the entire continent, covering it all in over a mile of water in under three hours. This wave would circumnavigate the globe several times, and it would take up to a week for the oceans to settle back down into their original resting places, having completely decimated and resurfaced the globe in the process

The book makes many other eyebrow-raising claims which run the gamut from an argument that the book of Genesis doesn't describe the creation of the planet, but rather the seven days that it took for the oceans to stop sloshing around and the skies to clear, to the argument that Jesus spent the missing years of his early life in India.

I really don't know what to make of any of this, but it does raise some interesting questions:

First of all, why did the C.I.A. classify this book in the first place? It's hard to know for sure. The general thrust of the book seems to be focused on these recurring geological cataclysms and their historical and spiritual tie to the history and development of humanity. Could this really be something that the C.I.A. was taking seriously?

Perhaps.

But the thing is—nothing about that idea was new. Even back when the book was originally supposed to be published in 1966 this idea wasn't new. Hapgood's book, *Earth's Shifting Crust*, had already been out for almost a decade. And in the decades since countless books have been written putting forward similar hypotheses, and none of those have been classified. So what was different about this book?

Second of all, why was the C.I.A. even paying attention to this? Chan Thomas was allegedly a former Air Force employee, but it's unclear—at least from my research—what his job was. However, he wasn't in the service. He had a degree in electrical engineering from Dartmouth, but he spent most of his career positioning himself as "the world's leading authority in the field of cataclysmic geology" and claimed to have accurately predicted a number of earthquakes. In fact, most of his message seemed to be apocalyptic, focusing on the idea that the last time this cataclysm supposedly occurred was 6500 years ago and that we're due for another one basically any minute.

But once again—none of this is new or different. Each new decade brings a new crop of New Age prophets who come bearing overly grandiose claims about their own abilities and dire apocalyptic predictions for the future of humanity. So once again, why do they care about this book and this author? And why risk legitimizing something by classifying it when they could have just let it slowly drift into obscurity, collecting dust in basements and used book stores? Why was this on their radar? Why pay attention to it at all?

I don't have those answers. And I in no way mean to suggest that this oddball little book that was declassified by the CIA proves anything about either the reality or the nature of cyclical, planet-wide cataclysms.

But it seems clear that the CIA, at least in the 1960s, saw this information as potentially being a threat to our national security—and *that* is interesting.

Alignment Of Megalithic Structures

But there's one more piece of evidence that, to be honest, blew my mind completely.

So, we know that megalithic structures were almost always built in alignment with the cardinal directions or significant astronomical events like the solstice. And, in many places, these alignments are done with a level of precision that humanity wouldn't be able to replicate for thousands of years.

But, strangely, this doesn't apply to all megalithic structures. In certain places, particularly in South America, Mesoamerica, and Egypt, the alignments are slightly off at certain sites. Why would these alignments be so important in some cases, but not in others?

A study was done by author and scientist Mark J. Carlotto that looked at 200 megalithic sites around the world and found that almost half of them couldn't be explained in terms of traditional alignments.

But remember our old friend Charles Hapgood and his theory that the poles shifted dramatically three times over the past 100,000 years? Carlotto hypothesized that the alignments of these sites might correspond to Hapgood's proposed pole locations.

Taking the group of unexplained sites, he developed an algorithm that would generate a best-fit model for the location of the poles they were aligned to, and the model is shockingly close to what Hapgood proposed.

Both Hapgood and Carlotto predict that the most recent pole before our current one was in Hudson Bay and are only 200 miles apart from each other. Carlotto predicted a pole in Greenland that was immediately preceded by a pole in the Norwegian Sea, and Hapgood's Iceland/Norway Pole is 1250 miles away from the former and 300 miles away from the latter. And finally, Carlotto predicted a pole in the Bering Strait that is about 1500 miles from Hapgood's pole in the Yukon. Not exact, but pretty close.

So, if we take for granted for a moment that Carlotto is correct and that the alignment of these megalithic structures confirms Hapgood's poles and his general thesis then we are able to attach a general time period to the structures that are aligned to each of these poles.

And when we do this, what do we find?

Well, not surprisingly, we find that the majority of the megalithic sites align to our current pole, including the Pyramids of Giza and the Delphi Amphitheater in Greece.

However, certain sites, such as Puma Punku, Tiwanaku, Baalbek, the Western Wall in Jerusalem, and Tenochtitlan in Mexico are aligned to the Greenland pole, which according to Hapgood's climate data would make them at least 50,000 years old—or around the same age as the sunken city in Cuba.

And shockingly, there are a number of sites that align to the Bering Sea pole. And, interestingly, a disproportionate amount of those sites are in Peru. Among them are the sites whose foundations of massively cut and precisely placed megaliths we discussed earlier that were built upon again and again, without ever coming close to matching their former glory—including Coricancha and The Temple of the Three Windows at Machu Picchu.

And these sites, which feature some of the most precisely cut megaliths known to us, may also be the oldest. If Hapgood and Carlotto's hypotheses are correct, these sites would be *at least* 100,000 years old.

Now listen, I find the fact that both Hapgood's climate data and the alignment of anomalous megalithic sites both point to the same general locations for these alternate pole sites to be pretty damn compelling. However, I'm not sure that I'm 100% willing to accept them.

It's clear that we've undershot the dawn of human civilization by a few thousand years. But 100,000 years? That presents a whole new set of problems and mysteries.

First of all, as far as we know, there shouldn't even have been humans in the Americas 100,000 years ago. It was previously estimated that humans first arrived in North America around 15,000 to 20,000 years ago by crossing a land bridge in the spot where the Bering Strait is now between modern day Russia and Alaska. And although recent findings have pushed that date back to 30,000 years ago, there's still a 70,000 year gap to account for.

And it's not just the timeline that presents an issue. There's also the location. There are few places in the Americas that you could go that would be farther than Peru is from the Bering Strait and still be in the Americas. So if—somehow—there was human civilization in Peru 100,000 years ago, it calls that entire migration story, where they started in the north and moved south, into question.

But if our human ancestors didn't walk across the land bridge to the Americas how would they have gotten to South America in the first place? By boat seems like the obvious answer, until you consider just how vast the Pacific Ocean is. The distance from Peru to Australia is 15,000 miles—or more than half of the circumference of the Earth. To say that that would be surprising would be a massive understatement.

And it wouldn't just throw the story of how humans came to the Americas into question, but also, everything we think we know about the migration of anatomically modern humans out of Africa and across the continents.

And even if we were somehow able to gather enough evidence to justify that dramatic rewrite of human history, we're still left with the most glaring and inescapable problem:

How would a high civilization capable of the astounding megalithic stonework that we see at these sites in Peru possibly have arisen 40,000 years before the earliest cave paintings that we've found? It just doesn't make sense. It's simply not possible.

But what if…

What if it wasn't a *human* civilization?

It's a wild thought, but it isn't a crazy one.

The admission by the Pentagon, the declassified videos, the rising bipartisan demand from Congress that they be given access to data surrounding what is clearly a credible national security threat, Tom Delonge's high-level government advisors that were confirmed by the Wikileaks Podesta emails, and the growing number of highly credible people from our military and intelligence agencies, as well as private aerospace companies, who are coming forward to say that *the UFO phenomenon is real and we need to take it seriously*—all of these things point to one stunning conclusion—**that there is a highly advanced, non-human presence on this planet *right now*.**

And that profound and, at this point, unavoidable fact means that it is entirely likely that it happened in the past, as well. And we should be willing to at least *consider* what that might mean for the story of humanity on this planet.

The Esoteric Tradition & UFOs

You know…or not. All of this is very fun and very interesting to talk about, and I haven't made a secret of the fact that I find the evidence to be quite compelling. But if you still aren't convinced, that's totally fine. I respect that.

But the good news is that, even if these arguments for non-human intervention in humanity's distant past have not swayed you, our time here was still not wasted. We'll actually need all of this to help lay the foundation for what's to come.

What these last three chapters have given you is the basic skeleton of the esoteric tradition. So what do I mean by that?

According to Google, "ancient esoteric tradition" is a modern scholarly term useful for designating currents in Hellenistic and Late Antique Mediterranean culture that are concerned with the mediation of some kind of absolute knowledge via a dialectic of secrecy, concealment, and revelation.

In more straightforward terms, there is a collection of beliefs that we tend to categorize as "occult" that are rooted in the belief in some special kind of ancient knowledge and ultimate truth that was entrusted to a select group of scholars and intellectuals. The ancient Egyptians—and then later the Greeks and the Romans—referred to these as the "mystery schools" and their initiates include some of the greatest thinkers of the ancient world like Aristotle, Socrates, and Plato. In our modern day world, we have the freemasons and other secretive societies that are the direct descendants of this tradition.

So you don't need to believe any of this—the mysterious origins of human civilization, guidance and revelation from the gods or some other non-human intelligence, or the idea that our ancestors had exposure to highly advanced ideas and technologies in our distant past.

But you do need to understand that, throughout the last several thousands years there have been many people who *do* believe something along those lines. And the search for that ancient knowledge has been the driving force behind some of the most significant events in human history from the Crusades to our exploration of space.

So that's where we'll pick up in book two—with the story of how a renewed interest in these ancient mysteries lead to a pagan spiritual awakening in Germany in the early 20th century, and subsequently, to two of the most important events not just of the last century, but in all of human history—the rise of Nazi Germany and the Moon landing.

And yes, UFOs and the mysterious intelligence behind them is at the center of everything that is to come.

I'll see you then.

The UFO Rabbit Hole: Book Two

Coming in early 2023.

Bibliography

Chapter 1

Brimelow, B. (2018, March 12). "What the f--- is that thing?": Mysterious video shows Navy pilots encountering UFO. *Insider.* https://www.businessinsider.com/video-navy-pilots-encountering-ufo-2018-3

CDC. (2020, February 11). *2019 novel coronavirus.* Centers for Disease Control and Prevention. https://www.cdc.gov/coronavirus/2019-ncov/index.html?CDC_AA_refVal=https%3A%2F%2Fwww.cdc.gov%2Fcoronavirus%2F2019-ncov%2Ffaq.html

Contributors to Wikimedia projects. (2022, August 22). *Sonic boom.* Wikipedia. https://en.wikipedia.org/wiki/Sonic_boom

Cooper, H., Blumenthal, R., & Kean, L. (2017, December 16). Glowing auras and 'black money': The Pentagon's mysterious U.F.O. program. *The New York Times.* https://www.nytimes.com/2017/12/16/us/politics/pentagon-program-ufo-harry-reid.html

Cooper, H., Blumenthal, R., & Kean, L. (2019, May 26). 'Wow, what is that?' Navy pilots report unexplained flying objects. *The New York Times.* https://www.nytimes.com/2019/05/26/us/politics/ufo-sightings-navy-pilots.html

Daugherty, G. (2019, May 16). When top gun pilots tangled with a baffling tic-tac-shaped UFO. *HISTORY.* https://www.history.com/news/uss-nimitz-2004-tic-tac-ufo-encounter

Einstein. (1971). *The evolution of physics.* Cambridge University Press.

Graves, G. (2021, October 12). Conspiracy theories are a danger to your health—and that's the truth. Here's why. *Prevention.* https://www.prevention.com/health/mental-health/a37898458/conspiracy-theories-harm-your-health/

Live, W. P. (2021, June 3). UFOs & national security with Luis Elizondo, former director, Advanced Aerospace Threat Identification

Program. *The Washington Post.*
https://www.washingtonpost.com/washington-post-
live/2021/06/08/ufos-national-security-with-luis-elizondo-former-
director-advanced-aerospace-threat-identification-program-aatip/

Robb, A. (2017, November 16). Rolling Stone. *Rolling Stone.*
https://www.rollingstone.com/feature/anatomy-of-a-fake-news-
scandal-125877/

Roose, K. (2021, October 22). What is qanon, the viral pro-trump
conspiracy theory? *The New York Times.*
https://www.nytimes.com/article/what-is-qanon.html

Strauss, D. (2020, April 28). Pentagon releases three UFO videos
taken by US navy pilots. *The Guardian.*
https://www.theguardian.com/world/2020/apr/27/pentagon-
releases-three-ufo-videos-taken-by-us-navy-pilots

Sullivan, G. D. & M. (2019, May 20). These 5 UFO traits, captured
on video by Navy fighters, defy explanation. *HISTORY.*
https://www.history.com/news/ufo-sightings-speed-appearance-
movement

To The Stars Academy of Arts & Science. (2017a). FLIR1: Official
UAP footage from the USG for public release [Video]. In *YouTube.*
https://youtu.be/6rWOtrke0HY

To The Stars Academy of Arts & Science. (2017b). Gimbal: The first
official UAP footage from the USG for public release [Video]. In
YouTube. https://www.youtube.com/watch?v=tf1uLwUTDA0

To The Stars Academy of Arts & Science. (2018). Go Fast: Official
USG footage of UAP for public release [Video]. In *YouTube.*
https://www.youtube.com/watch?v=wxVRg7LLaQA

Venosa, A. (2016, January 14). *Breaking point: How many g-forces can
humans tolerate?* Medical Daily.
https://www.medicaldaily.com/breaking-point-whats-strongest-g-
force-humans-can-tolerate-369246

Chapter 2

Becker, A. (2018). What is real?: The unfinished quest for the meaning of quantum physics. Basic Books.

Biography. (2017, April 28). Nicolaus Copernicus. Biography. https://www.biography.com/scientist/nicolaus-copernicus

Contributors to Wikimedia projects. (2022a, August 20). Lockheed U-2. Wikipedia. https://en.wikipedia.org/wiki/Lockheed_U-2#Cover_story

Contributors to Wikimedia projects. (2022b, August 21). Roswell incident. Wikipedia. https://en.wikipedia.org/wiki/Roswell_incident

Coulthart, R. (2021). In Plain Sight: An investigation into ufos and impossible science.

Einstein's greatest blunder? — Donald Goldsmith. (n.d.). Harvard University Press. Retrieved September 13, 2022, from https://www.hup.harvard.edu/catalog.php?isbn=9780674242425

Einstein's relativity and everyday life. (n.d.). PhysicsCentral. Retrieved September 13, 2022, from https://physicscentral.com/explore/writers/will.cfm

Garber, M. (2014, June 15). The man who introduced the world to flying saucers. The Atlantic. https://www.theatlantic.com/technology/archive/2014/06/the-man-who-introduced-the-world-to-flying-saucers/372732/

History.com Editors. (2019, December 2). Arms race. HISTORY. https://www.history.com/topics/cold-war/arms-race

Kotlikoff, L. (2017, December 8). Has our government spent $21 trillion of our money without telling us? Forbes. https://www.forbes.com/sites/kotlikoff/2017/12/08/has-our-government-spent-21-trillion-of-our-money-without-telling-us/?sh=37a725ba7aef

Missions Impossible: The Skunk Works Story. (2022, April 7). Lockheed Martin. https://www.lockheedmartin.com/en-us/news/features/history/skunk-works.html

Planetary motion: The history of an idea that launched the scientific revolution. (2009, July 7). https://earthobservatory.nasa.gov/features/OrbitsHistory

Podcast, T. U. (2021). That UFO Podcast - Episode 54 - Luis Elizondo - Listener questions part one [Video]. In YouTube. https://www.youtube.com/watch?v=N1DZMD4T2_I

Ptolemaic system. (n.d.). Encyclopedia Britannica. Retrieved September 13, 2022, from https://www.britannica.com/science/Ptolemaic-system

Roulette, J. (2020, October 5). Musk's SpaceX wins Pentagon award for missile tracking satellites. Reuters. https://www.reuters.com/article/us-space-exploration-spacex-satellites/musks-spacex-wins-pentagon-award-for-missile-tracking-satellites-idUSKBN26Q3A1

Skibba. (2018, March 27). Einstein, Bohr and the war over quantum theory. Nature. https://www.nature.com/articles/d41586-018-03793-2

The Twining Memo. (n.d.). Retrieved September 13, 2022, from http://www.roswellfiles.com/FOIA/twining.htm

Theories of Everything with Curt Jaimungal. (2021). Luis Elizondo on Biological UFO Samples, Remote Viewing, and explaining "Somber" #UFOamnesty [Video]. In YouTube. https://youtu.be/wULw64ZL1Bg

Thompson, A. (2021, May 24). The logic-defying double-slit experiment is even weirder than you thought. Popular Mechanics. https://www.popularmechanics.com/science/a22280/double-slit-experiment-even-weirder/

Thomson, L. (2019, May 30). The ten most expensive military aircraft ever built. Airforce Technology. https://www.airforce-technology.com/features/most-expensive-military-aircraft/

The UFO Rabbit Hole

Tillman, N. T., Bartels, M., & Dutfield, S. (2022, January 5). Einstein's theory of general relativity. Space. https://www.space.com/17661-theory-general-relativity.html

Vallee, J. (2014). Passport to Magonia: From folklore to flying saucers.

Webster, D. (2017, July 5). In 1947, A high-altitude balloon crash landed in Roswell. The aliens never left. Smithsonian Magazine. https://www.smithsonianmag.com/smithsonian-institution/in-1947-high-altitude-balloon-crash-landed-roswell-aliens-never-left-180963917/

Wilkins, J. (2020, November 18). The Nuremberg UFO Sighting of 1561 - Lessons from history - Medium. Lessons from History. https://medium.com/lessons-from-history/the-nuremberg-ufo-sighting-of-1561-4078ecfcd946

Chapter 3

Andrews, B. (2019, July 30). If wormholes exist, could we really travel through them? Discover Magazine. https://www.discovermagazine.com/the-sciences/if-wormholes-exist-could-we-really-travel-through-them

As many as six billion Earth-like planets in our galaxy, according to new estimates. (n.d.). ScienceDaily. Retrieved September 13, 2022, from https://www.sciencedaily.com/releases/2020/06/200616100831.htm

Baldwin, E. (2021, October 29). 10 strange animals in the mariana trench. Oceaninfo. https://oceaninfo.com/list/mariana-trench-animals/

Blythe Bernhard. (2013, December 16). Castlewood eating disorder lawsuit to be dismissed. STLtoday.Com. https://www.stltoday.com/lifestyles/health-med-fit/health/castlewood-eating-disorder-lawsuit-to-be-dismissed/article_28cf3275-a29e-5391-997e-243bdae361c3.html

Boeree, L. (2018, July 3). Fermi paradox: Why haven't we found aliens yet? Vox. https://www.vox.com/science-and-health/2018/7/3/17522810/aliens-fermi-paradox-drake-equation

Bostrom, N. (n.d.). Anthropic bias: Observation selection effects in science and philosophy. Notre Dame Philosophical Reviews. Retrieved September 13, 2022, from https://ndpr.nd.edu/reviews/anthropic-bias-observation-selection-effects-in-science-and-philosophy/

CARON, C. (2011, November 28). Therapist "brainwashed" woman into believing she was in satanic cult, attorney says. ABC News. https://abcnews.go.com/US/therapist-accused-implanting-satanic-memories/story?id=15043529

Cockell, C. S. (2018). The Equations of Life: How physics shapes evolution. Basic Books.

Conca, J. (2020, July 21). How extremophile bacteria living in nuclear reactors might help us make vaccines. Forbes. https://www.forbes.com/sites/jamesconca/2020/07/21/if-extremophile-bacteria-can-live-in-nuclear-reactors-maybe-they-can-help-us-make-vaccines/?sh=1a0f64dc7619

Contributors to Wikimedia projects. (2022a, August 7). List of alleged extraterrestrial beings. Wikipedia. https://en.wikipedia.org/wiki/List_of_alleged_extraterrestrial_beings

Contributors to Wikimedia projects. (2022b, September 2). Phoenix lights. Wikipedia. https://en.wikipedia.org/wiki/Phoenix_Lights

Creighton, J. (2014, July 19). The kardashev scale - Type I, II, III, IV & V civilization. Futurism. https://futurism.com/the-kardashev-scale-type-i-ii-iii-iv-v-civilization

Denis, D., French, C. C., Rowe, R., Zavos, H. M. S., Nolan, P. M., Parsons, M. J., & Gregory, A. M. (2015). A twin and molecular genetics study of sleep paralysis and associated factors. Journal of Sleep Research, 24(4). https://doi.org/10.1111/jsr.12282

Emslie, K. (2016, May 26). Awake in a Nightmare. The Atlantic. https://www.theatlantic.com/health/archive/2016/05/sleep-paralysis/484490/

Evidence for microbial life on Mars: Fossilized bacteria? (n.d.). American Museum of Natural History. Retrieved September 13, 2022, from https://www.amnh.org/learn-teach/curriculum-collections/cosmic-horizons-book/fossil-microbes-mars

Festival, W. S. (2019). What is Antimatter? [Video]. In YouTube. https://www.youtube.com/watch?v=ycmL-8GaA7Y

Geggel, L. (2021, July 6). A brief history of dinosaurs. Live Science. https://www.livescience.com/3945-history-dinosaurs.html

Greg. (2008, October 7). Michio Kaku - Impossible science. The Daily Grail. https://www.dailygrail.com/2008/10/michio-kaku-impossible-science/

Howell, E. (2021, December 17). Fermi Paradox: Where are the aliens? Space. https://www.space.com/25325-fermi-paradox.html

Kennedy, C., & Lau, A. (2021, June 30). Most Americans believe in intelligent life beyond Earth; few see UFOs as a major national security threat. Pew Research Center. https://www.pewresearch.org/fact-tank/2021/06/30/most-americans-believe-in-intelligent-life-beyond-earth-few-see-ufos-as-a-major-national-security-threat/

Lacina, L. (2019, December 13). How Betty and Barney Hill's alien abduction story defined the genre. HISTORY. https://www.history.com/news/first-alien-abduction-account-barney-betty-hill

Mann, A. (2019, August 1). What is a dyson sphere? Space. https://www.space.com/dyson-sphere.html

Merino, N., Aronson, H. S., Bojanova, D. P., Feyhl-Buska, J., Wong, M. L., Zhang, S., & Giovannelli, D. (2019). Living at the Extremes: Extremophiles and the Limits of Life in a Planetary Context. Frontiers in Microbiology, 0. https://doi.org/10.3389/fmicb.2019.00780

Pearce, J. M., & Denkenberger, D. C. (2018). A national pragmatic safety limit for nuclear weapon quantities. Safety, 4(2). https://doi.org/10.3390/safety4020025

Ph.D., R. V. (2012, November 4). Implanting false memories. Psychology Today. https://www.psychologytoday.com/us/blog/media-spotlight/201211/implanting-false-memories

Podcast, T. U. (2021). That UFO Podcast - Episode 27 - Luis Elizondo (video) [Video]. In YouTube. https://www.youtube.com/watch?v=ggyW_PuOcw8

principle of mediocrity. (n.d.). Encyclopedia Britannica. Retrieved September 13, 2022, from https://www.britannica.com/science/principle-of-mediocrity

Radio, F. T. B. (2021). Ep. 1371 FADE to BLACK Jimmy Church w/ Lue Elizondo□ : UAPs, TTSA, and the Future of Disclosure [Video]. In YouTube. https://www.youtube.com/watch?v=RrPYFQaJe3g&t=1s

Retherford, B. (2018, October 17). Extraterrestrials might look like us, says astrobiologist. Forbes. https://www.forbes.com/sites/billretherford/2018/10/17/extraterrestrials-might-look-like-us-says-astrobiologist/?sh=41989c594f0c

Scharf, C. A. (2021, February 21). Until recently, people accepted the 'fact' of aliens in the solar system. Scientific American. https://www.scientificamerican.com/article/until-recently-people-accepted-the-fact-of-aliens-in-the-solar-system/

Shanahan, J. (2018, August 7). These are the most likely places to harbor alien life in our solar system. Forbes. https://www.forbes.com/sites/jesseshanahan/2018/08/07/these-are-the-most-likely-places-to-harbor-alien-life-in-our-solar-system/?sh=69cc1e247004

Siegel, E. (2020, June 16). 36 alien civilizations in The Milky Way? The science behind A ridiculous headline. Forbes. https://www.forbes.com/sites/startswithabang/2020/06/16/36-alien-civilizations-in-the-milky-way-the-science-behind-a-ridiculous-headline/?sh=1ea839e879c5

The evolution of flight. (2016, December 21). Science World. https://www.scienceworld.ca/stories/evolution-flight/

Time, P. S. (2015). 5 REAL possibilities for interstellar travel [Video]. In YouTube. https://www.youtube.com/watch?v=EzZGPCyrpSU

What are the most Earth-like worlds we've found? (2021, August 23). The Planetary Society. https://www.planetary.org/articles/earth-like-worlds

Wood, C. (2019, November 1). What is convergent evolution? Live Science. https://www.livescience.com/convergent-evolution.html

Chapter 4

Coulthart, R. (2021). In Plain Sight: An investigation into UFOs and impossible science. HarperCollins.

David, L. (2020, January 20). Are the aliens us? UFOs may be piloted by time-traveling humans, book argues. Space. https://www.space.com/aliens-time-traveling-humans-ufo-hypothesis.html

Hanks, M. (2021, October 20). UFOs disabled weapons at nuclear facilities, according to these former USAF officers. The Debrief. https://thedebrief.org/ufos-disabled-weapons-at-nuclear-facilities-according-to-these-former-usaf-officers/

Hastings, R. L. (2008). UFOs and nukes: Extraordinary encounters at nuclear weapons sites.

Herper, M. (2013, June 7). No, this is not how the human face might look in 100,000 years. Forbes. https://www.forbes.com/sites/matthewherper/2013/06/07/no-this-is-not-how-the-human-face-might-look-in-100000-years/?sh=6e46e0602c78

Knox, P. (2021, October 16). I saw giant UFO disable 10 live nukes at top secret base – and Pentagon is covering it up, says US air f... The Sun. https://www.thesun.co.uk/news/worldnews/16430919/witnessed-giant-ufo-shut-down-10-nukes-secret-

The UFO Rabbit Hole

base/?utm_campaign=native_share&utm_source=sharebar_nati
ve&utm_medium=sharebar_native

Masters, Dr. M. P. (2019). Identified flying objects: A multidisciplinary scientific approach to the UFO phenomenon. Masters Creative .

Mysteries, L. L. (2012, September 19). Future humans will all look Brazilian, researcher says. Insider. https://www.businessinsider.com/future-humans-will-all-look-brazilian-2012-9

Neilson, S. (2020, September 30). Time travel is theoretically possible, new calculations show. But that doesn't mean you could change the past. Insider. https://www.businessinsider.com/time-travel-possible-changing-past-isnt-physics-says-2020-9

Olson, P. (2013, June 7). How the human face might look in 100,000 years. Forbes. https://www.forbes.com/sites/parmyolson/2013/06/07/how-the-human-face-might-look-in-100000-years/?sh=3c00a9705a96

Theories of Everything with Curt Jaimungal. (2021). Ross Coulthart on ufos, Wilson memo, SAFIRE project, and human abductions #nasatellthetruth [Video]. In YouTube. https://youtu.be/JM3kxeU_oDE

Thompson, A. (2016, September 29). Why does time go forward instead of backward? Popular Mechanics. https://www.popularmechanics.com/science/a23140/why-time-goes-forward/

Time travel is possible, but it's a one-way ticket – ScienceBorealis.ca Blog. (2020, November 16). Scienceborealis.ca Blog. https://blog.scienceborealis.ca/time-travel-is-possible-but-its-a-one-way-ticket/

Chapter 5

admin. (2021, May 7). How many Caves Are There in the World? Enter the Caves. https://enterthecaves.com/how-many-caves-are-there-in-the-world/

"Aliens and UFOs at world's deepest lake." (n.d.). Retrieved September 13, 2022, from https://siberiantimes.com/other/others/features/f0077-aliens-and-ufos-at-worlds-deepest-lake/

Axe, D. (2009, July 24). Russian navy declassifies Cold War close encounters. WIRED. https://www.wired.com/2009/07/russian-navy-declassifies-cold-war-close-encounters/

Contributors to Wikimedia projects. (2022, September 13). Jacques Vallée. Wikipedia. https://en.wikipedia.org/wiki/Jacques_Vall%C3%A9e

Effinger, G. A. (1973). Chains of the sea: Three original novellas of science fiction. X-S Books.

Gaia Hypothesis - An overview. (n.d.). ScienceDirect Topics. Retrieved September 13, 2022, from https://www.sciencedirect.com/topics/earth-and-planetary-sciences/gaia-hypothesis

Puiu, T. (2021, December 28). What is the world's deepest cave? ZME Science. https://www.zmescience.com/science/what-is-the-worlds-deepest-cave/

Ravindran, S. (2016, February 29). Inner earth is teeming with exotic forms of life. Smithsonian Magazine. https://www.smithsonianmag.com/science-nature/inner-earth-teeming-exotic-forms-life-180958243/

r/UFOs - A Strange Answer in Today's Delonge/Elizondo AMA. (n.d.). Reddit. Retrieved September 13, 2022, from https://www.reddit.com/r/UFOs/comments/hnaqyz/a_strange_answer_in_todays_delongeelizondo_ama/

Seeker. (2014). How many dimensions does the universe have? [Video]. In YouTube. https://www.youtube.com/watch?v=YoJ5BRghuAk

Seven species that used to be cryptids. (2020, December 12). ScIU. https://blogs.iu.edu/sciu/2020/12/12/seven-cryptids-species/

String theory. (n.d.-a). Symmetry Magazine. Retrieved September 13, 2022, from https://www.symmetrymagazine.org/article/may-2007/explain-it-in-60-seconds-string-theory

String theory. (n.d.-b). New Scientist. Retrieved September 13, 2022, from https://www.newscientist.com/definition/string-theory/

Sutter, P. (2021, August 23). What is multiverse theory? Live Science. https://www.livescience.com/multiverse

Think, B. (2011). Michio Kaku: The multiverse has 11 dimensions [Video]. In YouTube. https://www.youtube.com/watch?v=jI50HN0Kshg

TV, Q. (2020). Did The Soviet Union discover aliens in the deepest lake in the world? [Video]. In YouTube. https://www.youtube.com/watch?v=DVq7gBH70WE

Vallée, J. (1990). Five Arguments Against The Extraterrestrial Origin Of Unidentified Flying Objects. Journal of Scientific Exploration, 4, 105–117.

Vallee, J. (2014). Passport to Magonia: From folklore to flying saucers.

Which is greater? The number of atoms in the universe or the number of chess moves? (n.d.). National Museums Liverpool. Retrieved September 13, 2022, from https://www.liverpoolmuseums.org.uk/stories/which-greater-number-of-atoms-universe-or-number-of-chess-moves

Why the many-worlds interpretation has many problems. (2018, October 18). Quanta Magazine. https://www.quantamagazine.org/why-the-many-worlds-interpretation-of-quantum-mechanics-has-many-problems-20181018/

(N.d.). Retrieved September 13, 2022, from https://i.redd.it/y04416we7ci71.png

Chapter 6

atmoosphere. (2013). Blink-182 interview backstage in the 90's (sub español) [Video]. In YouTube. https://www.youtube.com/watch?v=uc1d25WgJZQ

Clips, J. (2019). Travis Barker on Tom Delonge's UFO fascination [Video]. In YouTube. https://www.youtube.com/watch?v=8nGbaD3xG2E

Contributors to Wikimedia projects. (n.d.). Angels & airwaves. Wikipedia. Retrieved September 13, 2022, from https://en.wikipedia.org/wiki/Angels_%26_Airwaves

Contributors to Wikimedia projects. (2022a, September 1). Tom DeLonge. Wikipedia. https://en.wikipedia.org/wiki/Tom_DeLonge

Contributors to Wikimedia projects. (2022b, September 12). Blink-182. Wikipedia. https://en.wikipedia.org/wiki/Blink-182

Cooper, H., Blumenthal, R., & Kean, L. (2017, December 16). Glowing auras and 'black money': The Pentagon's mysterious U.F.O. program. The New York Times. https://www.nytimes.com/2017/12/16/us/politics/pentagon-program-ufo-harry-reid.html

Doyle, P. (2016, April 27). Rolling Stone. Rolling Stone. https://www.rollingstone.com/music/music-news/inside-tom-delonges-ufo-obsession-blink-182-turmoil-50572/

Doyle, P. (2019, June 4). Rolling Stone. Rolling Stone. https://www.rollingstone.com/music/music-features/tom-delonge-interview-ufo-footage-angels-airwaves-blink-182-843812/

Koala, R. P. (2019). Tom Delonge's UFO Timeline part 1 [Video]. In YouTube. https://www.youtube.com/watch?v=4BjUK5V5sTg

Nero182. (2008). Tom talks about Blink 182 break up [Video]. In YouTube. https://www.youtube.com/watch?v=Il5VXjbQGNg

Newman, J. (2015, January 27). Rolling Stone. Rolling Stone. https://www.rollingstone.com/music/music-news/blink-182s-hoppus-barker-blast-ungrateful-disingenuous-tom-delonge-68807/

Radio, F. T. B. (2016). Ep. 515 FADE to BLACK Jimmy Church w/ Tom DeLonge☐: Sekret Machines☐: LIVE [Video]. In YouTube. https://youtu.be/VzLqBx5lN8Y

Re: A good read... - WikiLeaks. (n.d.). The Podesta Emails. Retrieved September 13, 2022, from https://wikileaks.org/podesta-emails/emailid/51537

Theories of Everything with Curt Jaimungal. (2021). Tom Delonge on CE5, skin walker, alien bodies, and Luis Elizondo's departure [Video]. In YouTube. https://www.youtube.com/watch?v=LVmxzCF-oeo

To the stars*. (n.d.). To The Stars*. Retrieved September 13, 2022, from https://tothestars.media/

To The Stars Academy of Arts & Science. (2017, October 11). To The Stars Academy Of Arts & Science launches today. PR Newswire. https://www.prnewswire.com/news-releases/to-the-stars-academy-of-arts--science-launches-today-300534912.html

Chapter 7

Clips, J. (2019). Travis Barker on Tom Delonge's UFO fascination [Video]. In YouTube. https://www.youtube.com/watch?v=8nGbaD3xG2E

Cooper, H., Blumenthal, R., & Kean, L. (2017, December 16). Glowing auras and 'black money': The Pentagon's mysterious U.F.O. program. The New York Times. https://www.nytimes.com/2017/12/16/us/politics/pentagon-program-ufo-harry-reid.html

DeLonge, T., & Levenda, P. (2017). Sekret Machines: Gods: An official investigation of the UFO phenomenon. To The Stars.

DeLonge, T., & Levenda, P. (2019). Sekret machines: Man: Sekret machines gods, man, and war. To The Stars.

Koala, R. P. (2019). Tom Delonge's UFO Timeline part 1 [Video]. In YouTube. https://www.youtube.com/watch?v=4BjUK5V5sTg

Patrick, D. (2021). Dr. Steven Greer - Talks ufos [Video]. In YouTube. https://www.youtube.com/watch?v=EhIITMb5pZk

Radio, F. T. B. (2016). Ep. 515 FADE to BLACK Jimmy Church w/ Tom DeLonge□ : Sekret Machines□ : LIVE [Video]. In YouTube. https://www.youtube.com/watch?v=VzLqBx5lN8Y&t=3088s

Rogan, J. (2017, October 26). #1029 - Tom DeLonge. Spotify. https://open.spotify.com/episode/2ybsXdWAtxqLBdRByLb2YG?si =Gwzyqx2sQ_eN87BMMDGBJg&nd=1

Theories of Everything with Curt Jaimungal. (2021a). Ross Coulthart on ufos, Wilson memo, SAFIRE project, and human abductions #nasatellthetruth [Video]. In YouTube. https://www.youtube.com/watch?v=JM3kxeU_oDE&t=7986s

Theories of Everything with Curt Jaimungal. (2021b). Tom Delonge on CE5, skin walker, alien bodies, and Luis Elizondo's departure [Video]. In YouTube. https://www.youtube.com/watch?v=LVmxzCF-oeo

Tom DeLonge. (n.d.). Coast to Coast AM. Retrieved September 14, 2022, from https://www.coasttocoastam.com/guest/delonge-tom-55135/

WikiLeaks. (n.d.). The Podesta Emails. Retrieved September 13, 2022, from https://wikileaks.org/podesta-emails/

Chapter 8

Carroll, A. E. (2018, November 5). Peer Review: The worst way to judge research, except for all the others. The New York Times. https://www.nytimes.com/2018/11/05/upshot/peer-review-the-worst-way-to-judge-research-except-for-all-the-others.html

Geologic activity. (n.d.). Mount Rushmore National Memorial (U.S. National Park Service). Retrieved September 14, 2022, from https://www.nps.gov/moru/learn/nature/geologicactivity.htm

Tarlach, G. (2016, June 1). Everything worth knowing about ... scientific dating methods. Discover Magazine. https://www.discovermagazine.com/planet-earth/everything-worth-knowing-about-scientific-dating-methods

Whitcomb, I. (2021, January 10). How do scientists figure out how old things are? Live Science. https://www.livescience.com/scientists-dating-methods.html

Chapter 9

Contributors to Wikimedia projects. (2022, May 3). V. Gordon Childe. Wikipedia. https://en.wikipedia.org/wiki/V._Gordon_Childe

Cremo, M. A., & Thompson, R. L. (1999). The hidden history of the human race. Torchlight Pub.

Curry, A. (2008, October 31). Gobekli Tepe: The world's first temple? Smithsonian Magazine. https://www.smithsonianmag.com/history/gobekli-tepe-the-worlds-first-temple-83613665/

Evans, P. H. R. (2019). The waterside ape: An alternative account of human evolution. CRC Press.

Firestone, R. B. (2019, July 24). Disappearance of ice age megafauna and the younger dryas impact. Capeia. https://beta.capeia.com/planetary-science/2019/06/03/disappearance-of-ice-age-megafauna-and-the-younger-dryas-impact

Fox, M. (2017, June 7). We're older than we thought: New find pushes human origin back 100,000 years. NBC News. https://www.nbcnews.com/news/world/we-re-older-we-thought-new-find-pushes-human-origin-n769376

Frank, A. (2018, April 13). Was there a civilization on Earth before humans? The Atlantic. https://www.theatlantic.com/science/archive/2018/04/are-we-earths-only-civilization/557180/

Schoch, R. M. (2021). Forgotten Civilization: New Discoveries on the Solar-Induced Dark Age. Simon and Schuster.

The Associated Press. (1992, February 9). Scholars dispute claim that sphinx is much older. The New York Times. https://www.nytimes.com/1992/02/09/us/scholars-dispute-claim-that-sphinx-is-much-older.html

ABOUT THE AUTHOR

Kelly is the host of the The UFO Rabbit Hole podcast. In her "real life" she's also a branding and marketing expert. Her work is centered on the role of storytelling as a catalyst for change and expanding human awareness. She is passionate about leveraging her skills and expertise to support the UFO disclosure movement and all of those engaged in the important work of driving forward our understanding of who we are as a species and where we fit in the great cosmic order of things. She lives in Ohio with her partner, George, and her cat, Chicken.

Printed in Great Britain
by Amazon